RECONNECTING PEOPLE AND WATER

Water management in industrialised western countries has long been seen as a technical process associated with pipes, drains and bureaucracies. This technical model of water management is now being questioned. This book examines the nature of contemporary water management and the prospects for and barriers to different forms of engagement with the public.

In particular, it shows how historical and social scientific understandings develop and question current water management norms in relation to water in the landscape, water in the home and the hidden management of water beneath our streets and behind our walls. It is shown that the four-fold challenges of climate change, urbanisation, changing environmental standards and fiscal accountability mean that we can no longer rely on unseen technical fixes to erase the threats of pollution, water shortages and floods. Such concerns offer two prompts for public engagement and participation. First, on a purely instrumental level, public engagement can complement, or offer an obvious alternative to, technical fixes. Second, public engagement may provide a route to find new ways of addressing water and related challenges.

The author offers a unique social science perspective on many of the socio-technical issues facing the management of water in urban settings in developed countries, where urban is interpreted broadly to include all areas served by piped water. Drawing on historical context and an extensive review of the published literature, as well as the author's own empirical studies, the work prompts broader discussions about how we manage water in contemporary society.

Liz Sharp is a Senior Lecturer in the Department of Urban Studies and Planning at the University of Sheffield, UK.

Earthscan Water Text Series

Reconnecting People and Water
Public Engagement and Sustainable Urban Water Management
By Liz Sharp

Key Concepts in Water Resource Management
A Review and Critical Evaluation
Edited by Jonathan Lautze

Contesting Hidden Waters
Conflict Resolution for Groundwater and Aquifers
By W. Todd Jarvis

Water Security
Principles, Perspectives and Practices
Edited by Bruce Lankford, Karen Bakker, Mark Zeitoun and Declan Conway

Water Ethics
A Values Approach to Solving the Water Crisis
By David Groenfeldt

The Right to Water
Politics, Governance and Social Struggles
Edited by Farhana Sultana and Alex Loftus

RECONNECTING PEOPLE AND WATER

Public Engagement and Sustainable Urban Water Management

Liz Sharp

LONDON AND NEW YORK

First published 2017
by Routledge
2 Park Square, Milton Park, Abingdon, Oxon OX14 4RN

and by Routledge
711 Third Avenue, New York, NY 10017

Routledge is an imprint of the Taylor & Francis Group, an informa business

British Library Cataloguing in Publication Data
A catalogue record for this book is available from the British Library

Library of Congress Cataloging in Publication Data
A catalog record for this book has been requested

ISBN: 978-0-415-72844-7 (hbk)
ISBN: 978-0-415-72845-4 (pbk)
ISBN: 978-1-315-85167-9 (ebk)

Typeset in Bembo
by Cenveo Publisher Services

CONTENTS

ACKNOWLEDGEMENTS

The writing of this book has involved structuring and ordering understandings that I have acquired during more than 15 years of research and teaching about water on the engineering–social science interface, and through my studies about the physical and human environmental interactions that preceded that. This means that there are some institutions and many people who have been important in the development of my ideas.

I want to acknowledge the universities of Bradford and Sheffield for supporting my career development and fostering my unusual interdisciplinary focus. I am also grateful to the Engineering and Physical Sciences Research Council and the European Union Framework 7, whose recognition of the need for social science to examine water issues has been crucial in supporting my work. Specifically, I would like to acknowledge the importance of the following grants:

- Water Cycle Management for New Developments (WaND, GR/S18373/01)
- Urban Rivers and Sustainable Living Environments (URSULA, EP/F007388/1)
- The Pennine Water Group Platform Grant (PWG, EP/1029346/1)
- Tailored Water Solutions for Positive Impact (TWENTY65, EP/NO10124/1)
- PREPARED EU Framework 7 Collaborative Project no. 244323

Some of the people who have been critical in helping me to understand the hydrology, engineering and ecology of water include Richard Ashley, Barney Lerner, Simon Tait, Joby Boxall, Vanessa Speight, David Butler, Paul Jeffrey, Lorraine Maltby, Dusi Thomas, Dominic Scott, Adrienne Walsh, Stuart Lane, Keith Richards, Adrian Saul, Phil Warren and Chris Jones.

The idea of the book grew from the 'Water and its Management' module I taught at the University of Bradford's Department of Geography and Environmental Science. It has therefore had the benefit of the many important and interesting questions that numerous students asked of me, and also of supportive colleagues including Nigel Copperthwaite, Sarah George, Carl Heron and Amy Dodds (née Walton).

Some of the social scientists who stimulated and helped to develop my understandings of the interactions between people and water include: Susie Molyneux-Hodgson, Peggy Haughton, Margi Bryant, Stephen Connelly, Christine Sefton, Cathy Knamiller, Emma Westling, Leona Skelton, Zoe Sofoulis, Ali Browne, Will Medd, Marta Rychlewski, Susan Owens and Paul Raven.

A big thank you to those who have read drafts or partial drafts of the book and provided comments, including: Emma Westling, Barney Lerner, Richard Ashley, John Flint, Chad Staddon, Nikky Wilson (who also helped with indexing), Ruth Levene, Phiala Mehring, Stephen Connelly and, among these, a special big thank you goes to Zoe Sofoulis (who helped me feel enthusiastic about the book before I wrote it, and then read and commented on the whole book in draft) and also Leona Skelton (who guided me in forming historical commentaries, and also copy-read an almost-finished version).

Finally, I would like to thank those who have helped me personally through the process of producing this book, including Zoe Sofoulis, Richard Ashley, Vanessa Speight, Barney Lerner and Leona Skelton (all for encouraging me), Martin Hoffman (in particular, for feeding me!) and to my sister and parents for their love, and for first enabling my interest in water to develop, aged four, when building dams on a stream in Wales.

INTRODUCTION

What does it mean to reconnect with water?

In contemporary western societies, water is almost invisible. It is vital to our daily lives, yet so readily obtained and so quickly disposed of that we give it little thought. Rain may prompt a change of attire, but, except in extreme events, it impinges little on our lives. In many respects the invisibility of water is a symptom of the 'success' of water management. We have long delegated the roles of obtaining clean water, of draining our streets and of disposing of dirty water to experts who are so good at their job that we don't even see them doing it. In this respect, we are disconnected from the systems that make our everyday life possible.

'Good', a reader might say. Water professionals do their jobs and we are able to get on with our busy lives. We really do not want to go back in time and 'connect with water' through collecting our own bucketful from the well. What is the problem?

The answer is that achieving 'invisible water' for those who currently enjoy it comes at a great financial and environmental cost. Even more importantly, providing 'invisible water' for everyone would be impossible. Even in developed countries, where invisible water has been achieved for most, it is going to be extremely hard to maintain: this is because the pipes are old, the climate is changing and we expect our water systems to contribute positively to improving our environment.

Finding ways to 'connect' people more closely with water is one alternative to water remaining invisible that may be both cheaper and environmentally more sustainable. At its heart, people being more connected with water means the public taking a greater part in both decisions and action to achieve better water management. In effect, it means moving away from the 'one size fits all' model of water management that has dominated the way that water systems have developed over the past century. It means tailoring water management processes to suit local conditions and needs – both in terms of water and in terms of culture. It means local water systems being more actively managed by the public, in conjunction with their utility, whether by modifying their fresh water consumption, being more selective about what they dispose of, collecting and using rainwater for themselves, or possibly by accepting, owning or maintaining local drainage features or river banks. Reconnecting people with water means more complicated physical systems, combining centralised elements with decentralised elements under the control of the water user. And it means the business of water utilities moves from being almost entirely technical and operational to combining technical and operational elements with communication and negotiation with many different 'network managers'; that is, with stakeholders and members of the public managing elements of the water networks for themselves.

Reconnecting with water does not mean reverting to some pre-industrial nightmare in which each household takes full and absolute responsibility for their water supply and disposal. It does, however, mean learning lessons from the past and from other places so that we can find new ways in which members of the public, alongside utilities and other organisations, take appropriate responsibility for ensuring that local water systems are effective and sustainable. It is likely to mean that people develop a greater understanding about how water works within their locality. Drawing on Linton's (2008) understanding of how the hydrological cycle has been replaced by a 'hydrosocial cycle', *Reconnecting People with Water* could be called 'moving towards hydrosocial water management'.

Central arguments of *Reconnecting People with Water*

So, 'reconnecting with water' means the public understanding water management and challenges in their locality and being active participants in the process of addressing those challenges. But what, precisely, am I saying about such 'reconnecting'? In practice, there are four interconnected arguments that I make through this book.

First, I make a descriptive argument about the past. As implied by the title *Reconnecting People with Water*, I am suggesting that in the past many people had a part to play in making decisions and taking actions that contributed to the collective process of managing water and waste water in their locality. Through having to think and know about water in their locality, I suggest, we all had a connection with the weather, the geography and other people's cultural expectations that is lacking today. Through the gradual development of what I call 'technocratic water management' I suggest that these connections have been severed. Reaching its zenith in the industrialised world in the 1970s and 1980s, technocratic water management promised a 'modern' utopia in which people did not need to engage with their environment. This first descriptive argument that there has been a gradual process of disconnecting people from water is made explicitly in Chapter 1; it then discusses details of these processes in relation to water supply (Chapters 3 and 4), water quality management (Chapter 6) and flood management (Chapters 7 and 8).

Second, I make an argument about the state of contemporary water management. It is, of course, the case that – even at its most technocratic and disconnected extreme – water management has always operated in a social context. However, the difference between water management now and in the mid- to late twentieth century is that many water utilities and local amenity groups are seeking to explicitly find ways to reconnect people with water. 'Hydrosocial water management' is already present in some industrialised countries, I suggest, because water management organisations are reaching out to work with their customers, local residents, environmental groups and others to manage water systems differently. They are asking for water-saving behaviour, care with the use of drainage systems, attention and action to improve local waterways, and seeking volunteers for water-related duties such as being river wardens. Moreover, hydrosocial water management has also arisen because some local citizens, amenity or environmental groups are themselves prompting changes to water management activities as a means to improve the role of water in their locality. This second argument, that there is a change underway in water management, is made throughout the book through numerous examples of where and how 'reconnecting' practices are already undertaken.

Third, I make a values-based argument that more high-quality 'reconnecting' is required. Specifically, I suggest that high-quality hydrosocial water management promises (1) water services that are equally effective and better value for money than those currently available, (2) co-benefits in other areas like nature conservation and landscape beauty and (3) enhanced democracy, with public debate and decision-making about water, the landscape and our collective management thereof, both symbolising and mediating citizens' attachments to their localities and communities. In terms of water management, I suggest that while many technical solutions will continue to be relevant, more knowledge by water experts of public and lay perspectives could help in the design and selection of tailored solutions, and specifically that changes in behaviours or expectations can be both financially and environmentally less costly than traditional technical approaches. In terms of joining water management up with other services, more knowledge of water management by different publics and infrastructure service providers could offer the potential for integrating water decision-making more closely into other decisions about the management of our landscape and our wider investments – for example, in parks and in economic regeneration. In terms of democracy, the concern is that, in the past, technical specialists have made decisions about new water technologies or investments with limited reference to public perspectives. As citizens, we have interests in how our society delivers collective services and maintains our environment and hence public perspectives are relevant and important in water decision-making that forms part of a democratic process.

This values-based argument about the need for more 'reconnecting' is made through many illustrative examples about reconnecting with water in contemporary water management. I've already noted above that these examples are used to show the changing nature of water management, but by assessing and evaluating the experiences discussed, the same examples can offer normative guidance about whether and how reconnecting can work. By showing where, how, for whom and why reconnecting has worked (and, indeed, where, how, for whom and why it has not worked), the book begins to help examine the nature of 'high-quality' hydrosocial water management. It illustrates the advantages and disadvantages of new arrangements for different stakeholders, with the aim of helping readers think through arrangements that might work in their context.

It is also useful to note at this point that there are important critiques that might be made about whether *reconnecting with water* can really make a difference within the modern world. In arguing about how we need to reconnect people with water it is important to both explore and address these perspectives. I will touch on these critiques in Chapter 1, but then explore precisely what they say and how they might be answered more fully in the Conclusion, Chapter 10.

Fourth, I argue that the consequence of the need for more reconnection between the public and water is that a whole new set of skills and knowledge is needed to inform our water management institutions and processes. Specifically, they need some new types of professionals *within* the utilities, but they also need to draw on a new knowledge from universities, academies and institutes that is *external* to the utility. To begin with the internal argument, if water utilities are to successfully engage the public they need professionals with an understanding of public communication and public participation as well as water services. Employing such professionals is likely to enable utilities to develop and deploy a far more nuanced and differentiated understanding of the different 'publics' with whom they are working, and for whom different types of interactive communications need to be developed and tailored. Employing such professionals is also likely to occur alongside a broader shift of the 'knowledge ecologies' of water utilities (Fam *et al.*, 2015) to embrace ways of understanding and knowing that include the contextual and the particular, as well as the conventionally generalisable and statistical. Hence, moving on to the 'external argument', as well as drawing on technical specialists to help develop and innovate in their physical systems, water utilities will also draw on social scientists of water governance, institutional behaviour and water practices and equally on historians of particular localities and times.

The value of context-specific knowledge is argued and illustrated in a variety of ways. In Chapter 1 I make an academic argument about the need for such knowledge. Through many of the subsequent chapters I have used a historical narrative to help readers to understand how aspects of our water systems have evolved. These can be seen as illustrating the strength and importance of historical knowledge by example. Likewise, specific cases illustrating when and how people have been 'reconnected with water' are themselves narratives that can be remembered and can inform subsequent practice. Finally, but importantly, on a series of occasions I draw on social science theory to show a form of generalisation that is not statistical, but can nevertheless inform research and practice. Such generalisations take the form of frameworks or checklists that constitute social science theory. Distilling previous scholarship, these frameworks help observers step outside the specific details of a case and to consider what alternative approaches are possible. Through direct argument, through illustrative cases and through theory, therefore, *Reconnecting People with Water* argues for the importance and value of knowledge drawn from the interpretive social sciences and humanities to help understand and support future water management.

Drawing these four arguments together, the aim of this book is to explore how we are reconnecting with water and how this could happen more and better. In more formal terms, it will develop and articulate the nature of a participatory, social science-informed hydrosocial water management and hence support practitioners and others in building a more socially, economically and environmentally sustainable water system. The focus is less on the technical processes of this transition, but rather on how public cultures might shift to incorporate water management, on the one hand, and how water utility cultures might shift to incorporate enhanced interactions with their stakeholders, on the other.

What is new?

Considered in isolation, each of the four arguments discussed in the last section has been articulated before, at least in some form or another (see Table I.1). In this

TABLE I.1 Predecessors in relation to each of the four arguments made in this book

Argument	*Prominent predecessors in making this argument*
People have become disconnected from water through past water management	Passionately and convincingly argued in Strang's (2004) anthropology of water management. Also apparent from Karen Bakker's (2004, 2014) economically oriented accounts of water history, and from Jamie Linton's discussions of the development of the hydrological cycle (2008, 2014)
People and water management are now becoming better connected	Most particularly argued by Linton (2014) and by Linton and Budds' (2014) discussions about hydrosocial renewal. Claimed in general terms by those promoting integrated water resource management, ecosystems services and adaptive water management (see references in Chapter 1). Also made in relation to specific sectors – for example, in relation to flooding, see Johnson and colleagues (2007) and Nye and colleagues (2011)
More high-quality hydrosocial water management is needed	Stated, but implications for water management not fully developed, by geographers Linton and Budds (2014). Part of calls for water sensitive cities (e.g. Wong and Brown, 2008). Implicit in critiques of water management such as made by Molle (2008). Made by commentators within the water sector, such as Murphy (2014). Partly parallels calls for high-quality public participation in other sectors such as land-use planning (Arnstein, 1969; Healey, 1997) and environmental management (Reed, 2008)
Water management needs new skills and knowledge drawn from social science and humanities	Championed by new social science researchers of water, such as Chappells and Medd (2008), Doron and colleagues (2011), Sharp and colleagues (2011b) and Fam and colleagues (2015), as well as by innovative natural science practitioners, such as Lane (see Lane *et al.*, 2011). Parallels calls for more integration between social science, natural science and policy, such as Chilvers and Evans (2009)

respect, the book has important influences and precursors. But these precursors do not take away from the unique contribution that is being offered.

This book makes an original contribution by bringing these arguments together. In particular, it seeks to bridge a yawning gap between social and political commentaries, on the one hand (arguments 1 and 2), and engineering/ecological accounts/aspirations, on the other (argument 3). Critical historical and political commentators such as Bakker (2004) and Linton (2014) highlight issues of power, particularly stressing those people or interests that have lost out through the treatment of water as a commodity. In contrast, those writing from engineering and ecological specialisms tend to stress new insights, processes and technologies that enable water management to be improved (see discussions about integrated water resources management, ecosystem services and adaptive water management in Chapter 1). The first set of work is historically situated, with implied or explicit criticism about the present. The second is better positioned in terms of understanding the day-to-day activities of water managers, but optimistically assumes that different interests can be aligned and solutions found. By applying their background in social science and the humanities to the real challenges of water management, a small set of 'social' scholars of water are already, constantly, trying to bridge this gap in small ways within specific projects and papers (argument 4). But there is only so much that can be said in a conference presentation, report or paper. This book offers a more studied and considered narrative, seeking to illustrate and exemplify the myriad ways that interpretive social science has something to say about water management, and how, through interpretive understandings, new forms of people-sensitive practice can be imagined, developed and enhanced. Through this narrative, I hope this book will establish a solid foundation from which more interpretive social science of water can be developed, and hence to help build a new and better form of hydrosocial water management.

It is useful to point out a couple of additional unique or unusual features that are the consequence of this original contribution.

First, and somewhat surprisingly, *Reconnecting People with Water* is unusual in bringing together explorations about different parts of the water sector in one volume. Rather than focusing on the practice of water supply management, water quality control, flood risk management, or water's role in the landscape in isolation, here they are considered together. (This said, the volume does not consider every aspect of water management – for example, the use of water for navigation and the connections between fresh water and marine management are not considered.) It is important and useful to discuss these different parts of the water sector together because they interact very closely, and, in particular, proposals for the reform of any one of these 'sectors' has profound implications for others.

Second, the book is a perhaps rather daring attempt to synthesise material between disciplines. Specifically, it combines water engineering understandings and knowledge with historical accounts alongside social science case studies and theory. As I explain below, I am not an expert in all of these subjects, but having worked at the boundaries between these areas for a number of years I believe I have a unique and interesting perspective that this book aims to share. Altogether, I hope by articulating and exemplifying exciting changes that are underway in the water sector I can contribute to shaping and furthering the development of hydrosocial water management.

What does *Reconnecting People with Water* cover?

The book has an ambitious scope, which is to explore changes in water management – including water supply, drainage and flood management – that are occurring in the industrialised world. Even more ambitiously, it seeks to do so in a historical context, and seeks to present some social science ideas through which such changes can be understood. But how can one volume possibly make robust arguments that cover such an enormous scope?

A premise of hydrosocial water management is that history and context matter. Hence, while the four arguments outlined in the previous section offer the big storyline, they are illustrated through a string of particular and peculiar local narratives. This is not a compromise: to systematically sample the entire industrialised world would be overwhelming. The book's examples are drawn from a variety of localities. While my base in the north of England has yielded many cases, elsewhere in the UK, continental Europe, North America and the Antipodes also contribute examples to the narrative. Even where one chapter unapologetically focuses on one locality through a variety of local stories, understandings are contextualised and supported in two ways. First, using literature to position in the historical and international literature enables examples to be understood in terms of their relative degree of difference/innovation. Second, use of theory provides kernels of succinct summated knowledge from elsewhere, prompting specific questions and suggesting interpretations of any case presented. My aim is that the reader comes out with some understanding of the richness and variety of practice, but also armed with critical questions about context and about the relative roles of experts and the public through which to read and understand water management practices in contexts they know. In summary, then, the robustness of this volume's arguments is achieved through the geographical and historical range of these examples, the coherence of the narrative in which they are framed and their positioning in theoretical, historical and geographical context.

The focus on the industrialised world requires clarification. The term 'the industrialised world' refers to the richer countries with longer established processes of urbanisation and industrialisation that would previously have been called 'developed' and that social scientists euphemistically describe as 'the Global North'. Because the stories told position these nations in the context of their industrialisation and urbanisation (as well as subsequent de-industrialisation), it is possible or indeed likely that some of the stories have wider resonance and may hold lessons for cities undergoing rapid industrialisation and urbanisation today. However, I am not a scholar of emerging economies (i.e. the nations which might have previously been described as the 'developing' world) and it is for others to explore and clarify how far the trends and issues discussed herein have wider meanings for cities beyond the industrialised world.

Finally, it is important to position the arguments made in disciplinary terms. By doing so I hope not only to make the material more understandable from different perspectives, but also to provide transparency about my position, hence enabling readers to assess whether and how my interpretations and understandings will be similar or different to their own.

I write as an 'environmental social scientist' who has worked closely with academic and practitioner water engineers and ecologists for 15 years. My perspective is positioned in the 'interpretive policy analysis' tradition (Hajer and Wagenaar, 2003; Yanow and Schwartz-Shea, 2006; Fischer, 2009). This means I am interested in governance in terms of what is said, how it is understood and what is done – whether this 'doing' is by the government, the private sector, or by individuals. I am centrally concerned with the questions: 'what values are perpetuated through how things are done; how is this changing and, how should it change?' Like others writing in the interpretive policy analysis tradition, I favour more extensive participation by the public in governance, but would suggest that this has to be developed in a careful and critical manner. The position is broadly optimistic in believing that things could and should be done better than they have been in the past. As well as exploring interpretive analyses of water management, I also draw on water research developed within related social science traditions – most notably political ecology – and 'practices' research from science and technology studies. Not only do these traditions provide very useful and important research on water, they also broadly share a critical and interpretive stance. I also reach outside the social sciences, drawing particularly on history and to some extent on engineering. (In practice a lot of the engineering I include is drawn from researching and working with engineers, rather than from engineering publications.) Although my understandings of these areas are partial, the research is drawn on selectively and as part of the broader argument. Indeed, I am grateful for the support of critical readers who have worked in these traditions and have helped me to position this work appropriately.

Who is the book for?

This can be seen as a 'fusion' book, seeking to introduce new ideas to two distinct communities.

One principal target audience for this book is technically oriented professionals and students of water management. These people, with backgrounds ranging from civil engineering through environmental science to aquatic ecology, are increasingly being asked to interact with the public about decisions that are being made. By introducing this audience to some of the complexities of interacting with the public (and the wide range of people that this phrase encompasses), the book can guide their interactions and potentially structure their decisions about when and how to access more specialised social science support.

A second target audience is social science-trained students and professionals who are new to water systems. This audience, I suggest, have skills and knowledge which may be of relevance in a new people-oriented water management, but need to understand enough of how water systems work to be able to engage intelligently. Introducing this audience to water systems and their working, including the routes through which the current systems have evolved, is another objective.

So, why address the audiences together: why not write two books? An important point is that this book is written from a social science perspective, albeit in the context of many years of work with technical academics and professionals working on water issues. Technically trained people who choose to read this book are already reaching out and looking for ways to work more closely with the public and other stakeholders within their water management practice. Through engaging with the issues covered in the book, these technically trained people will start to have a more social perspective. Likewise, those with a social science background who choose to read this book will start to understand something more about how engineering and ecological understandings have contributed to the current water system. The social knowledge and background is of relevance, but it needs to be translated into clear practical terms to be useful for the water professionals. In short, the book seeks to provide the means and background through which these two audiences can interact more easily. Hence, *Reconnecting People with Water* could additionally be called 'connecting technical water management with social science'.

On this basis, the book seeks to provide:

- an introduction to the physical and institutional structures through which water is currently managed, including insight on how the current systems evolved;
- examples of how water utilities and the public are working, and have worked, together differently, both through different cultures and through different times;
- an introduction to some of the concepts and research processes which might support a more people-oriented water management.

There are two important implications of seeking to appeal to two specialist audiences. First, it has been necessary to write in a way that can be understood across disciplines. Insofar as technical language is used I have sought to explain it. This means that, in addition to the two specialist audiences noted above, the material is accessible to an interested lay audience. Second, it is likely that not all parts of the book will feel relevant to all readers at all times. Some parts of what follows will be very familiar to some readers, while other parts may contain too much detail for individuals' contemporary concerns. Overall, while the book can be engaged with cover to cover, it is expected that many readers will start by dipping in for empirical examples, and then perhaps later feel the need to understand the concepts through which the research described was carried out.

Structure

In preparing this book my overall strategy has been to weave together two strands of reasoning. An empirical strand concerns the different 'parts' of water management. Broadly, I begin upstream and at the large scale and the book travels downstream and to smaller scales. The goal is not to make the reader a water manager, but to broadly understand how the different parts of the current system have evolved and fit together. The second theoretical strand of reasoning uses each chapter's empirical material as an illustration in the discussion of a relevant aspect of theoretical social science argument or theory. Likewise, here, the goal is not to make the reader a social scientist, but rather to illustrate the range of perspectives and questions that are opened up through viewing water management via an interpretive social science lens. These two – empirical and theoretical – strands are generally twisted together, demonstrating the usefulness and applicability of the social science for understanding the water situation.

Chapter 1, 'Visions for water management', describes changing 'visions' for urban water management through time, introducing the idea of hydrosocial water management as one concept among several that seeks to point to the future for water management. The chapter also introduces interpretive social science as demonstrating the stance on knowledge that is required for hydrosocial water management to function. In Chapter 2, 'Urban water use in context', the purpose is to contextualise urban water management. Changing ideas about the water cycle and changing conceptualisations of water resources across the world are presented. The chapter demonstrates the close link between science and practice; the concepts through which water supply is understood reflect and perpetuate changing emphases and concerns about the location and distribution of water resources. In Chapter 3 'The governance of water supply and demand', the focus is on the processes of water supply and how this has shifted through time. The chapter also discusses ideas about governance and the shifting understandings about the functions and geographical scales over which water supply should be integrated. In Chapter 4, 'Water in the home: learning from the past', the practices perspective is introduced. This is demonstrated through telling the stories of the changing practices through which we have consumed water in the home. The practices perspective is further elaborated in Chapter 5, 'Understanding water practices and mobilising change'. This chapter charts changing perspectives on how water providers influence the water demand of their customers, concluding that the practice perspective offers some potentially more inclusive and customer-centric ways to influence demand in the future. In Chapter 6, 'Water qualities', the emphasis shifts from water supply to water quality in both the tap and the river. Two contemporary case studies show action and compromises relating to water management today. The chapter also includes a three-part historical account of how London's sewer system developed as it did. The social science theme of the chapter concerns the extent to which the physical systems need to be centralised or decentralised. Chapters 7, 'Water out of place', and 8, 'Flood risk governance', chart the stories four different examples of localities vulnerable to flooding. In Chapter 7 the cases are described and put in the context of a wider shift from flood defence to flood risk management. Chapter 8 introduces Mary Douglas' cultural theory as a means to identify 'governance cultures'. This theory is then used as a means to understand the different cases presented in Chapter 8. In Chapter 9, 'Water in the landscape', the purpose is to explore some examples of how people have engaged with their local landscapes. The examples also show changing attitudes to the public and the normalisation of public mobilisation and participation in landscape-related decision-making. The material is drawn together in Chapter 10, the Conclusion, which focuses on what contribution the book has made to hydrosocial water management, why and how interpretive social science contributes to a hydrosocial approach and what implications the book has for water management.

1

VISIONS FOR WATER MANAGEMENT

Introduction

This chapter explores ideas about 'good' water management in cities, and how these have changed through time. Visions of good water management are very important not only because they guide what is designed and built, but also because they influence what institutions are formed for managing the water, and what water-use practices are expected of and communicated to the public. Because infrastructures, institutions and practices can take a long time to change, what is created in one era continues to influence what happens on the ground for a long time afterwards. Past visions of the future can therefore help us to understand our current water system. Meanwhile, articulating and developing our current picture of the future has the potential to influence future water management.

My key argument in this chapter is that a past vision of the future – which I call 'technocratic water management' – has profoundly influenced the water systems that we enjoy today. Though there are many alternative visions of water management in the contemporary era, I suggest that, at present, these give insufficient attention to the social and political side of water management. I therefore draw on and develop these ideas to voice my own vision: I articulate a case for a more socially aware and participatory approach to water that I call 'hydrosocial water management'.

Technocratic water management

Urban water management in the industrialised world has long been seen as a technical process associated with pipes and bureaucracies. In the idealised 'technocratic' picture of the modern city, people enjoy the benefit of standardised water amenities in their homes, factories and offices largely unaware of the technical and institutional infrastructure that provides clean water to properties, removes and treats waste water and manages rainwater, transporting it rapidly from roofs and roads. Nature is infinitely manageable, so lawns are green, floods are nasty memories of a bygone era and fish swim in the rivers.

The most visible and tangible manifestations of technocratic water management are large dams. Awe-inspiring structures that transform rural landscapes into sites for water and energy storage, dams can be seen as symbolising and, indeed, embodying human management and control over nature. In practice, the visible dams are just one end of a multi-tendrilled network of water supply pipes, treatment works and sewers, largely hidden from view and responsible for holding, pumping and processing to keep flows of water circulating through our cities. As well as these physical manifestations of technocratic water management, the system's operation also relies on a web of organisations and human expectations, ensuring that water bills are charged and paid, and that water services are used in particular ways – for example, that toxic paint is not poured down the drain. In the context of twenty-first-century visions of the future centred on self-propelling vehicles and other accoutrements of the smart-city, it may seem odd to understand these dams and pipes as one-time symbols of the future. However, as recently as the 1970s, dam construction was still seen as enabling industrial expansion and economic growth in the industrialised world. And although large dam projects now rightly attract worldwide controversy, they continue to be seen as crucial symbols of development in many emerging economies (International Rivers, 2016).

The development and implementation of technocratic water management has been closely associated with urbanisation. Typically, as towns grow water systems change in at least four ways (with huge variations in the intensity, chronology and scale of these changes, depending for whom, when and where they occur). First, the geographical impact of the water system spreads as the town 'catches' water from more distant locales, and also disposes of water further downstream. Second, the technologies of water systems become more sophisticated, entailing mass transport over greater distances, and increasing treatment through time. Third, water systems cease to be addressed on an individual or household level and, instead, become a communal problem which is centrally addressed by the town, or some section of the town, through delegating responsibility to an increasingly diverse set of experts working in different water-related (supply, drainage, sewage) fields. And, finally, household water systems become more standardised, and expectations about unlimited piped cold and hot water, waste water disposal and fully drained public areas become ironed into habitual behaviours.

These various changes to the water system reveal the assumptions underlying the technocratic model. Piped water supply, toilets and bathrooms are seen as markers of decency and civilisation. Lifestyles are expected to improve through time, and hence water demand is expected to rise. From the technocratic perspective, the needs of agriculture, industry and public consumers of water must be predicted and supplied; the 'technocratic city' is therefore assumed to sit within a relatively abundant natural world, which can provide as much water as is needed, with a large capacity for absorbing waste water. Money and the ingenuity of expert water managers are assumed to be able to find ways to address any physical or environmental problems that occur in the continuation of this model.

More broadly, what is here characterised as the 'technocratic model' parallels what Linton has called 'modern water' (2014) and what Karen Bakker has labelled the 'state-hydraulic paradigm' (2004). For Linton, a critical marker of the emergence of modern water was the shift from an understanding of many different waters that were fit for different uses and purposes, and with varied social, cultural and religious meanings, to a perspective focusing on a single uniform abstract totality 'water', to be developed and exploited by the modern state (Linton, 2014). For Bakker (2004), water supply and sanitation serve as material emblems of citizenship in the 'hydraulic state'. Market failure – for example, water's status as a public good, externalities related to water abstraction and disposal, and the natural monopoly status of water supply and drainage – provides the reason for state control over these resources. In line with the political ecology approach adopted by both theorists, water is understood to have become a resource that is exploited for political and economic power.

The power and the 'success' of the technocratic model of water infrastructure should not be underestimated. The technocratic model drove the development of water infrastructure in the industrialised world from the early twentieth century until at least the 1980s, and continues to drive new infrastructure in urbanising cities across the globe today. It has provided the means to deliver potable water to millions of households; it has also developed processes of dealing with sewage and with surface or storm water which, if not brilliant for the wider environment of our rivers and estuaries, has largely eliminated waterborne disease and ensured pleasant and healthy surroundings within our homes and streets. Hence, at the time of its development, and during large periods of its implementation, the technocratic model both promised and delivered significant societal benefits.

Questioning the technocratic model

The technocratic model's impact on the individual household is interesting. Clearly, potable water supply, flood defence and waste water disposal has saved the arduous labour or costs of transporting water, reduced health risk, ensured a pleasant urban environment and reduced vulnerability to extreme weather events. However, there are also costs, including a reduction in the variety of households' and communities' knowledge of water technologies and systems, especially of their understandings of local water systems' temporal and geographical patterns. Moreover, whereas households' immediate dependence on the weather has reduced, the less positive corollary is that their knowledge of how to protect themselves from extreme events has also been reduced, if not eliminated. Instead, we have all learnt to believe that 'the system' will protect us from most extremes. It is good in many respects because it means we feel secure. If, by any chance, the water system is challenged through flood or drought, other systems like emergency services, insurance and social services will ensure a speedy recovery. In some respects the fantastic all-providing technocratic system has infantilised us (the public); we are provided for, but also rendered dependent, incapable and unknowing. It is the technocratic system itself that has rendered water invisible.

The political ecology perspective would additionally stress that the impacts of the technocratic water system are differentiated according to people's wealth and power (Swyndegouw, 2004; Kaika, 2006; Moore, 2011). In both developed and developing economies, it is the poorest people who often pay the highest financial costs for water management. Poorer homes are also differentially located in flood prone areas, while the environmental costs of a degraded water system (for example, polluted waterways and smelly sewage works) are seldom experienced in the middle-class parts of town.

A number of factors have brought the technocratic model into question over recent decades. In this respect, as well as 'failing' to maintain community engagement and to work towards equality, the technocratic model is also 'failing' in relation to its own technical measures of success. First, rising environmental expectations mean that we have become increasingly aware of how our water systems are damaging or changing the natural environment. More extensive development, more stringent environmental regulations and denser populations mean that the technocratic model's assumption of an abundant natural world can no longer be operationalised. Second, urbanisation, globalisation and changing consumer lifestyles have meant that the load on our water systems has continued to increase. Particular challenges arise in specific locations – for example, attempts to live 'western' lifestyles in arid climates, leading to demands on very distant water sources or non-renewable groundwater resources. The spread of impermeable surfaces associated with urbanisation has also increased the water volume and pollution load in our drainage systems. Third, most western water systems include city-centre elements that were constructed more than 100 years ago. While this infrastructure has often weathered well, its location means that mending or expanding it is both expensive and disruptive. Fourth, however water management is financed, in most countries there is a concern about it costing too much, so water utilities must make a very convincing case to be permitted to invest in new infrastructure. Moreover, whereas nineteenth-century investment in water infrastructure was 'sexy' because it related to a vision of the modern society and drew public attention and excitement, extending this infrastructure to poor parts of town, or reinvesting and mending old infrastructure, has no such appeal. And, finally, the fifth, huge and uncertain wild card is climate change. Climate change means that we can no longer assume 'stationarity'; in other words, we cannot rely on future weather patterns emulating those of the recent past. Forecasters suggest that in many locations drought and flood events that are currently designated 'extreme' will become more normal, challenging all of the safety margins that are built into the water systems. Moreover, investment in technical measures to address these changes would make extensive use of carbon-demanding infrastructure, hence further contributing to climate change.

Together, environmental standards, urbanisation, ageing infrastructure, fiscal stringency and climate change seem to make a 'perfect storm' of pressures, suggesting that our technocratic model of water systems cannot survive in its current form.

What is at stake is the hydrological cycle's implicit understanding that humans can be separated from the waters they are seeking to manage.

Instead, as each of the challenges listed in the previous paragraph makes clear, and as Linton (2014) argues, people change waters, and also waters change societies. The technocratic understandings of modern water managed by a hydraulic state must therefore be replaced by a new 'hydrosocial cycle' in which the mutual influences between humans and water are acknowledged through two sets of changes. On the one hand, it needs to be recognised that our habits and expectations are shaped by the water infrastructure available to us; on the other hand, through concerted and collective effort we need to rethink our hydrosocial system to ensure that equality and sustainability are 'designed in' to future water management.

What does a 'hydrosocial' perspective mean for city residents of developed or emerging economies? While nobody is proposing that the advantages of the technocratic model be completely removed, a question is raised over whether and how we can maintain selected positive aspects of the technocratic model through building a new layer of resilience and robustness by individual, household and enterprise responsibility? How can key aspects of modern living be retained while moving on from the infantilised public with which the technocratic model is associated? This book addresses these very important questions.

New models for water management

During the last three decades, progressive water managers have proposed concepts, policies and practices to address perceived failures of the technocratic model. Examining these concepts, policies and practices enables an understanding of whether and how we have moved beyond the technocratic model. The discussion below therefore begins by describing three significant 'water concepts' rising to prominence over the past 30 years, highlighting their implicit understanding of the public, and linking to critiques in the literature. Having addressed water management concepts, the following subsection selects three instances of water management policies and practices to illustrate the same issues and their implications for water management on the ground.

Before embarking on this comparison, it is important to note the extent to which technocratic water management is and is not a 'model'. Unlike the concepts discussed below, no scientist or engineer ever articulated what I call 'technocratic water management', claiming that this approach provided the basis for hydraulic planning in the future. The technocratic approach is, however, a 'model' insofar as it is the extrapolation of trends that existed on the ground. It is also a 'model' because it is an ideal type rather than a reality; in other words, it is an extreme and totalised version of something that we have already, to a large extent. It should be noted, however, that technocratic water management has never fully been implemented because, as Sofoulis *et al.* (2005), Knamiller (2011) and Sefton (2008) have noted, there has always been some variety of practices within and between locations; and whatever the view of the majority, people retain the ability to find ingenious and individual ways to address extremes of water shortage or flood.

Water management concepts

So how do new water management concepts address the inadequacies of the technocratic model? In particular, how do they understand urban water management and the role of the public therein? Here, we review three important concepts that have influenced policy and practice over the past few decades.

With its disciplinary base in hydrology and water resource management, integrated water resource management (IWRM) rose to prominence in the 1990s, and was closely associated with the 1992 Rio conference on sustainable development and the linked Dublin principles on water management (Biswas, 2004; Savenije and Van der Zaag, 2008). IWRM seeks to promote 'the co-ordinated development and management of water, land and related resources, in order to maximise the resultant economic and social welfare in an equitable manner without compromising the sustainability of vital ecosystems' (GWP–TAC, 2000: 22). The emphasis on coordination of water and land resources highlights a frustration with the siloed decision-making and 'dis-integration' that arose from the technocratic model (Medema *et al.*, 2008). IWRM therefore criticises the way that the technocratic system separated water management into a series of increasingly specialised technical areas (for example, water supply management, water quality management and flood risk management), and hence failed to exploit and recognise the links between areas and the connected nature of the water system. IWRM also recognises that water is scarce (Dublin principle 1), and it should be treated as an economic resource (Dublin principle 4), to be shared following public participation (Dublin principle 2) (Jeffrey and Gearey, 2006). In this respect there is some acknowledgement of environmental problems of continuously increasing demand, and the implication that this can be addressed through economics and public participation. However, this process of 'raising the profile of the human factor in Hydrology' is recognised to require going outside the 'comfort zone' of geophysical sciences and engineering (Savinjie and Van der Zaag, 2008: 296). Cases in which organisations have explicitly sought to apply IWRM are often at a large sub-national or international 'river basin' scale, with a particular emphasis on the development of the right institutions, at multiple scales, to ensure that the appropriate range of stakeholders participate in decision-making (GWP, 2013; Medema *et al.*, 2008). A variation of IWRM, integrated urban water management (Mitchell, 2006), applies IWRM at a city scale.

Drawing on ecology and economics, ecosystems services (ES) seeks to define the benefits that people obtain from ecosystems so that the latter can be more fully accounted for within decision-making (Costanza *et al.*, 1997). For the ES approach, the environment is 'natural capital' employed to yield benefits, that may be of four types: provisioning (e.g. providing food), regulating (e.g. air circulation), supporting (e.g. soil formation) and cultural (e.g. recreation) (Fisher *et al.*, 2009). Some of these benefits may be synergistic – for example, a landscape providing flood mitigation may also offer leisure opportunities. But other benefits can be in tension: a landscape might offer either food growing or the regulating benefits of biodiversity conservation, but the same land cannot be fully used for both benefits simultaneously.

ES analyses identify the various ecosystem services associated with particular locations or environmental goods, and then estimate the trade-offs and synergies between different options for change, hence making the intangible environmental and social implications of decisions more explicit. ES studies are not confined to water but seek to address the whole range of natural resources. Many ES applications put an absolute financial value on ecosystem services; but some others focus on the trade-offs between different sets of services within specific scenarios, and hence to look at the relative values of different land-use choices (Cook and Spray, 2012). Alongside detailed study of the ecosystems at specific locations, stakeholder surveys and meetings derive the relative or absolute 'values' to be ascribed to different sets of ES.

Originating from ecological science the concept of 'adaptive management' and its water-related variant, adaptive water management (AWM), concentrate on the non-linearity and complexity of water systems (Folke, 2006; Pahl-Wostl, 2007; Medema *et al.*, 2008). Adaptive management aims to provide strategies for collective experimentation and learning in which institutions steer complex systems towards achieving the 'best' resilient state (e.g. Folke, 2006); in other words, the state that can be recovered quickly and completely after it has been shocked, and which can also continue to deliver required services in the context of gradual changes. There is great emphasis on the need for a dynamic process of monitoring and adjustment of the system, so that learning is obtained from 'experimental' interventions, which might be social or cultural (Pahl-Wostl, 2007; Medema *et al.*, 2008). Such adaptive approaches become possible in the context of a coordinated water strategy incorporating and mobilising the knowledge, values and practices of a variety of local participants and stakeholders (Pahl-Wostl *et al.*, 2008; Rouillard *et al.*, 2013; Murphy, 2014). Adaptive institutions therefore draw on both technical and social information about changes and their impacts. Many works on adaptation focus on the challenges of coordinating actions between different organisations and community groups in large socio-ecological systems (e.g. Olsson *et al.* 2004, Tompkins and Adger, 2004). However, adaptive management has been applied in a range of contexts, including within an organisation (Westling *et al.*, 2014) and at a city scale (Van der Steen and Howe, 2009).

The three concepts of integrated water resource management, ecosystems services and adaptive water management can all be seen as attempts to develop and improve the technocratic model; not so much 'revolution' as 'evolution', therefore. The concepts all critique and seek to develop different aspects of technocratic water management, reflecting their origin in various disciplines and relating to different professional fields. Only IWRM related specifically to addressing the challenges of water management. ES and AWM both came out of ecology and the management of natural landscapes, and hence are closely linked to issues about water, but also to biodiversity and other ecological concerns. The key focus of ES is to highlight the multiple values of ecosystems (and hence to ensure that they are sufficiently valued in the world). In contrast, adaptive management is more concerned with the interactions of those managing socio-ecological landscapes. In the recent literature the ES approach remains largely distinct, whereas for a range of scholars IWRM and AWM are increasingly talked about in one breath (e.g. Medema *et al.*, 2008).

Clearly, one significant difference between the technocratic model and these new concepts for water management is that the former has been implemented in many locations to a significant extent, while the latter have not, or only in a few locations to a very limited and debatable extent. This is not surprising. As noted above, what I am calling the 'technocratic model' is derived from its implementation. In contrast, IWRM, ES and AWM can all be seen as 'nirvana concepts' (Molle, 2008: 132) – that is, 'ideal image[s] of what the world should tend to' that 'present an attractive and useful focal point' but are open to widely different interpretations. As the term 'nirvana concepts' implies, the futures conjured by these concepts form hollow rallying cries, exciting a feeling of consensus but spurring diverse activities. In presenting and trying to sell an 'ideal future' which is different from the technocratic model, Molle suggests, advocates of nirvana concepts paint the future in broad positive terms with which few people would disagree. From Molle's perspective, then, it is not surprising that the concepts inspire, nor that they are criticised for their unclear definition (Medema *et al.*, 2008; Cook and Spray, 2012) and for the lack of evidence about benefits deriving from their implementation (Medema *et al.*, 2008; Norgaard, 2010; Cook and Spray, 2012). Notwithstanding these shortcomings, these different water management concepts are very useful for showing us how water management ideas and thinking have developed. In particular, the features that these concepts have in common with the technocratic model indicate which aspects of the technocratic approach endure in contemporary water aspirations, while those which differ from the technocratic model demonstrate which aspects of the latter are being rejected.

One notable difference between all of these approaches and the technocratic model is that the environment is given much greater and more explicit emphasis. IWRM starts from the recognition that water resources are limited, while ES and AWM emphasise the need to quantify and understand ecological processes. This is not surprising as environmental concerns are now widely acknowledged to be of great importance across the political spectrum, and in both developed and developing worlds.

A second notable change, of great relevance here, concerns the role of the public in decision-making. Whereas the technocratic model treated the public as passive beneficiaries of water services whose most active role was as a customer paying a bill, all of these models seek to bring public and stakeholder voices into decisions, either by directly involving them in decision meetings (e.g. Van der Steen and Howe, 2009), or by incorporating expressed economic preferences into the process of making choices (e.g. Costanza *et al.*, 1997). This change seems to demonstrate a recognition that the nature of water systems should not just be formed by the engineers who design them and the politicians who pay (or permit charging) for them. In this sense, the approaches begin to take on the idea that water and humans cannot be completely separated.

A feature that all three approaches have in common with the technocratic model is the prominent role given to science and knowledge in decision-making. The technocratic model lauded technical knowledge about the storage, conveyance and cleaning of water. Various different aspects of knowledge are additionally highlighted as important for the implementation of these concepts: for IWRM this concerns basin hydrology (Biswas, 2004); for ES it is ecological information about the different services and their 'value' (Cook and Spray, 2012); and for AWM the emphasis is on using different disciplines' skills to measure ecological, physical and social reactions to shocks and changes (Medema *et al.*, 2008). Given the right science, all four approaches suggest, better decisions will be made. The assumption that the right science will lead to the right policy-making is a common one for commentators seeking to ensure better governance. However, studies of how science and policy interact in practice suggest that the relationship between research (science) and policy are complex (Flyvbjerg, 2001; Jasanoff, 2003; Owens, 2005). From these perspectives the governance we experience is the interplay of dynamic relationships between stakeholders, knowledge and the existing infrastructure. Moreover, science itself may not present just one viewpoint, and should not be expected to deliver the 'answers' (Oreskes, 2004). We can also identify that some of the science to be drawn upon (basin hydrology, economics) is as profoundly committed to the separation of humans from their environment as the traditional engineering associated with the technocratic model.

A second notable enduring feature is the technocratic model's optimistic belief that a 'best' route forward can be identified and followed. Through drawing together the learning of different sciences and stakeholders, it is suggested, an optimised or most resilient set of choices can be identified (Fisher *et al.*, 2009; Folke, 2006). As a number of commentators have pointed out, however, an emphasis on science and expert decision-making may run in direct contrast to paying attention to the preferences of the public and, in particular, to those of people who are disadvantaged and without power (Jeffrey and Geary, 2006; Cook and Spray, 2012). Likewise, the expectation that scientists from different disciplines can reach an agreement about appropriate routes for action may be optimistic and, in particular, the integration of social sciences with natural and engineering science is frequently lacking (Biswas, 2005; Norgaard, 2010). Moreover, some of the participating stakeholders are also likely to have different priorities – for example, emphasising the perspective of 'the public' as customers, as citizens, or as encompassing a marginalised minority (Biswas, 2005; Molle, 2008; Medema *et al.*, 2008). Indeed, with the optimistic aspiration that a consensual route forward can be reached, each concept may be inadvertently compliant with the continued domination of those who hold power (Molle, 2008; Voß and Bornemann, 2011). In this respect, Voß and Bornemann's searing critique of adaptive management as a dangerously depoliticised view of governance might be equally applied to approaches derived from ES and IWRM.

> AM [adaptive management] appears as an approach that conceptually blocks out the nasty and dark sides of the political that may disturb rational problem solving.... The very effort of blocking out politics, however, renders AM itself a highly political concept as it tends to tacitly build on and reproduce given power relations.
>
> (*Voß and Bornemann, 2011: 9*)

In conclusion, this review of new models for water management shows a significant development from the technocratic model through the stress they lay on environmental considerations and public participation. However, the models give enduring emphasis to the role of science, implicitly positing a somewhat unrealistic picture of governance processes, with limited exploration as to how the interaction between public preferences and sciences might play out. In terms of implications for the public, we can conclude that, though the public are considered part of future decision-making, there are significant variations between and within approaches about which public(s) should be participating and how their participation feeds into decision-making. A particular tension can be noted between approaches that seek to draw on the public in a participatory manner and those that seek to represent their views through economics. There is also a question about how differing perspectives are reconciled, and whether the aspiration for consensus blocks out relevant science, and/or is likely to conceal the re-exclusion of the already disempowered.

Water management practice

If new water management *concepts* include other stakeholders, albeit inadequately, can the same be said for contemporary water management *practice*? Four examples are given below to illustrate how water managers are working with the stakeholders, and some of the challenges they face.

Mass communication about flooding in the Netherlands

With a quarter of the country below sea level, and much of its other land vulnerable to river flooding, the Dutch have a long tradition of cooperating in the management of flood risk from water through locally elected water boards – but how does this translate to the relationship between water authorities and the public today? Heems and Kothuis (2012) and Kothuis (2016) have explored the history of public communication about flooding in the recent Dutch context. Drawing on documentary analysis and interviews with water professionals as well as other stakeholders, they identify three different 'discourses' (or 'ways of talking about') about flooding. First, following the 1953 floods, they argue that a discourse of 'fight' sought to battle against the strong power of nature. It was on the basis of this battle with nature that a large 'Deltaworks' programme of new defences against the sea was justified. These large technical developments were seen subsequently, for a while, as having conquered the challenges of water management and a discourse of 'victory' prevailed. (Even the Dutch term for flood management, '*water safety*', conveys something of this discourse.) More recently, the challenges of climate change, and the need to pay attention to river and surface water management as well as flooding from the sea, have led to the new discourse of 'threat'. This is based on the idea that complete safety can never be achieved and that there will always be a need to prepare for the risk of catastrophic events. Analysing contemporary public communications about water safety in the Netherlands, Heems and Kothuis conclude that these tend to mix up the different discourses, creating confusing messages for the public. In particular, they highlight that whereas many water professionals speak and think in terms of the 'threat' discourse, the public are still oriented towards the discourse of victory and see no need to alter their behaviour. Communications that include both the victory and the threat discourse (e.g. that say 'we are very safe, but we can never be completely safe') tend to be heard for just their 'victory' message and the 'threat' part of the communication is ignored.

Local communication about water meters in south-east England

The background to the second case is the peculiar method of paying for water in England, by which some customers pay a fixed charge related to the value of their house, while others pay a variable charge linked to the amount of water they use. The water utility concerned was one of the English companies under the most pressure in terms of water supplies. Due to the lack of spare water capacity, in 2008, the water utility embarked on a process of piloting the compulsory metering of some of its customers in one part of the area. This process was based on the assumption that customers would use less water if they were metered, and hence that extra water supply could be won through demand management.

The utility selected a pilot town, largely because the town had one supply pipe that could be easily monitored. The utility set about writing to the residents, warning them about the change in the way they charged for water among the 50 per cent of residents who were not already metered. In addition to informing residents about meter installation, they also provided information about saving water to help residents respond appropriately to the change of charging system. The utility then began the process of meter installation. There was limited explicit opposition to this change in the way local people were charged for water, in the form of discussion in the town council. However, research with residents later revealed several misconceptions that created significant unnecessary bad feeling about the water utility (Knamiller and Sharp, 2009; Knamiller, 2011). First, many residents assumed that their town's selection for this exercise would mean a permanent difference between the way that they paid for water and the way that other people in the locality paid for water. In other words, they had not understood that installing meters in the town was a pilot process for compulsory metering that was later to be rolled out across the whole of the area served by the utility. In practice, this lack of understanding was not surprising, as the utility's communications did not mention this broader context for the trial. This first misunderstanding linked with the second misconception, which concerned the reasons for their town being selected as the location for the pilot. Residents of the pilot town had long felt that their town was treated as the poor relation of the larger neighbouring settlement. Upon hearing about the compulsory metering pilot, many residents believed that, further to what they saw as the normal pattern, the smaller pilot town had been picked on by people from the big town.

A subsequent stage in the process involved dividing the metered area up into four parts, and experimenting with different tariffs in each segment of the town. The research suggested that this process would not improve residents' perspective about the water company. Residents already perceived that they were non-consenting parties to a metering experiment that was being imposed on them. Effectively this second stage in the pilot project again treated residents as 'guinea pigs' in a second experiment, this time concerning tariffs. This type of social experimentation features little recognition that behaviours are not formed in isolation, but depend on perceived fairness compared to the treatment of others.

Public meeting in Yorkshire

The third example involves the UK Environment Agency (EA), and its concern to provide improved water habitats in the River Don in South Yorkshire (Westling, 2012). Here the EA was intending to employ environmental consultants to enhance the ecological habitat of a section of the river near the village of Fishlake. Having made their plans, public comments prompted the EA and their consultants to hold a public meeting to inform the residents about the changes and to answer questions about their concerns. The focus of the meeting was on giving information about the planned changes, though a questionnaire provided some opportunity for attendees to have an input into the scheme. The questionnaires asked respondents to express a preference from a menu of landscape elements, including better-maintained footpaths, enhanced access in other respects, information boards and a designated fishing area. In practice, the meeting did not run as the organisers anticipated. Difficult questions betrayed public hostility to the scheme and suspicions about the motives of the EA. The timing of the meeting – just a few days before the scheduled works – compounded the residents' impression that their interests were not important to the scheme design. Overall, the timing, aim and structure of the meeting did not successfully exploit the opportunity to strengthen the sense of ownership or support for the scheme among local residents.

Adapting to climate change in Wales

From the early 2000s, Dyr Cymru Welsh Water (DCWW) innovated to address flooding and water quality issues through the ambitious objective to remove all rainwater from sewers over a 30-year period: a significantly more ambitious goal than many of the other UK water companies. This goal has been implemented through a series of strategies, the most recent of which is the 'Rainscape' programme (DCWW, 2016). Linked to this stream of activity, DCWW collaborated with academics in the PREPARED European research project on adaptation to climate change. The PREPARED project in DCWW brought the asset management team and other teams together in a series of workshops to explore and develop the surface water management strategy and activities. The workshops are interesting because what the team liked about them reveals something of the limitations of day–to-day strategy-making and practice in a water utility. Specifically, by working together the teams were able to identify, name and (to some extent) identify solutions for problems that could not be addressed by one part of the organisation working alone. For example, the teams identified the need to coordinate better internally so that a joined-up face could be presented to external partners. They also pinpointed many areas for investigation and potential development. For example, one of the outcomes of the workshop recommended that the organisations should 'Examine how DCWW can justify paying for benefits which are not direct DCWW responsibility' (Rychlewski *et al.*, 2013). Overall, the workshops highlighted how the divisions of responsibilities associated with technocratic water management are hard to overcome. Even when stakeholders in the same organisation share a belief in the need for more integrated and adaptive management, lining up the goals of individual parts of an organisation to work in a reformed direction is very challenging and takes work beyond and outside of water companies' employees' day-to-day jobs (Westling *et al.*, 2014).

Common themes

What the examples above have in common is that they all involve water managers seeking to reach out and to work with the public or their stakeholders in different ways. In a sense, they all provide examples of water managers trying to work with stakeholders in the way that the new models of water management call for.

It is important to consider what motivated the interactions with stakeholders in these examples. In all but the third (public meeting in Yorkshire) example, public involvement was needed because the water professionals sought to mobilise stakeholder action; in these cases, this action was needed variously to achieve greater public flood resilience and to increase water conservation. Another motivation for stakeholder engagement was to inform people about forthcoming changes: for example, regarding the way they are charged for water or the setting of their local river. Only in the final case was stakeholder involvement motivated by a desire to understand more about the stakeholders' opinions and ideas, albeit focused on the 'internal' stakeholders working for the organisation. As Chapters 5 and 9 will demonstrate, mobilising public action, providing information and collating information about stakeholder views in order to shape water management actions are three of the important factors that motivate more stakeholder engagement. As is discussed below, none of them exactly constitutes 'public participation' in the sense of public input to important public decisions.

The examples also demonstrate certain difficulties in achieving the communication with stakeholders that is desired. Some of the issues that occurred include:

- communicating mixed messages
- not understanding or communicating locally important information
- a focus on individuals in isolation rather than a recognition that they form part of a community

- not listening to stakeholders' (i.e. employees', residents', or customers') knowledge and concerns
- insufficient attention to the variety of stakeholders.

The examples raise an important question about how water managers can effectively work with the public and other stakeholders.

Managing divergent views

The discussion of water management concepts and practice both demonstrate a contemporary struggle within water management. Traditional natural science and engineering science provides clear instructions about what should be done in certain situations: 'given an objective to improve the ecology, this is action you should take', or 'you want the pipe to last, you need to make it out of this material'. Seeking to involve other disciplines, stakeholders and the public can upset this clarity. A confusing variety of objectives are accompanied by seemingly 'irrational' considerations like politics, personal preferences or by another scientific discipline with different assumptions and a different rationality. Broadly, there are three potential ways for water professionals to respond to these potential contradictions, finding a route forward in the face of divergent views.

The first is the traditional response, which is to maintain the ultimate superiority of science over the other considerations. Like the technocratic model, this involves holding science as the ultimate arbiter in society and advocating the perspective suggested by the dominant science. If this approach were applied to the question of river restoration in Yorkshire, above, the EA would ignore the findings from the public meeting and implement the plan they made for the river in its original form. Water management would be split into those parts ruled by engineering science and those ruled by ecological science.

The second method accepts that science(s) needs to be viewed alongside other ways of seeing the world, but argues that the comparison needs to be weighed scientifically. Within this approach, money, or some other common metric, is used to explore the perceived costs and benefits of different routes forward. This is the approach that lies behind the technique called 'cost–benefit analysis' which was widely drawn on in the 1970s and 1980s (Pearce and Barbier, 2000). The same methods of assigning monetary values to nature are involved in the use of financially informed ES analyses, such as that by Costanza *et al.* (1997). If a financial ES approach were applied to the question of river restoration in Yorkshire, then the public's preference for a view of the river, and for lack of vegetation, would be given a monetary value that would then be balanced against another monetary value assigned to how restoration might improve the river's ecology. A decision would involve comparing the relative costs and benefits of developing the scheme or not.

The second method splits the process of the science (understanding causality) from that of assigning values. The challenging part of this method is working out what values to assign to phenomena like 'a pleasant river bank' that are not normally talked about in monetary terms. While economists have methods for interacting with the public to assign these values, these methods are very controversial and have received significant criticism (Flyvbjerg, 2001; Foster, 2003). A key argument of those opposing the monetary valuation of natural phenomena is the idea that knowledge and values cannot be separated; in other words, the knowledge and methods through which 'values' are investigated is imbued with a particular way of seeing and valuing the world. It seems that while the second method accepts that the values inherent in the science may not be the only values that are important, it uses another science (economics) to ascribe what they are and to determine what actions follow. Hence, in the end, a 'scientific' way of understanding the world is imposed.

The third method also recognises that there are a variety of different value systems in the world, but it does not seek to translate different value systems into a common metric. From this perspective, all ways of understanding the world (scientific or otherwise) are imbued with particular ideas about what is important to understand, measure and explain, and what is not. Rather than seeking to assign weight to different preferences, it advocates engaged discussion in which the different approaches to the issue can be understood, different responses considered and compromises made. Crucially, it advocates that arguments from all sides are subject to scrutiny, with questions asked and concerns raised from other perspectives. The challenge here is to find a way in which preferences can be voiced, choices discussed and concerns explored, while ensuring that excluded voices are able to be heard, but still leading in the end to decisions and action.

Such an engaged approach to decision-making could be read as one interpretation of what IWRM, ES and AWM are advocating. For example, IWRM's Dublin principle 3 specifically advocates that women – a previously much excluded group – are part of decision-making processes. However, as noted above, the vagueness and optimism of all three 'nirvana concepts' provides something of a fig leaf behind which 'business-as-usual' can continue.

A key issue relating to engaged decision-making is the weight that is given to scientific opinion. If scientists and policy-makers enter a decision-making forum with an attitude that the scientific answer is the only correct route forward, then it will inhibit their ability to listen. A more productive discussion can be achieved if all (different) scientists, water professionals, the public and other stakeholders are equally able to present their views and preferences. This approach recognises that particular scientific (or other) ways of measuring and understanding the world are already imbued with ideas about what is important – the social scientists would say that this means that each form of knowledge includes some implicit values. This perspective argues that while scientists are experts in their own field, local people are experts in their own locality, and stakeholders in their organisation – they all hold 'contributory expertise' (Healey, 1997; Collins and Evans, 2002). Moreover, decisions about the public interest – for example, whether a technology should be implemented or not – go beyond contributory expertise to require public debate about the best interests of society (Fischer, 2009: 160). To take part in participatory decision-making, therefore, scientists and water professionals need to embrace humility and to see their perspective as one valid perspective among many (Jasanoff, 2003). Clearly, they need to be respectful to their own integrity and to advocate their views with passion. However, they also need to be respectful to other people, to understand their perspectives and the values and understandings from which they are derived. Such a position can be a challenge for some professionals if their technical training has implied that science provides the single correct answer.

Of the three approaches to making decisions with stakeholders discussed above, it appears that only the third engaged approach avoids imposing a scientific perspective on stakeholders or the public. Unlike the imposition of scientific views associated with the technocratic model (the first method above), or the translation of all views into a common metric through economic valuation (the second method discussed above), participatory policy-making has the potential to be respectful to a variety of viewpoints.

Engaged water management: applying the hydrosocial model

What does it mean to achieve a process of water management that is respectful to a variety of viewpoints? Or, in Linton's terms, what does it mean to apply the hydrosocial model in practice?

The argument here is that it involves two interlinked changes in water management. First, it means a change in the way of interacting with stakeholders and the public that is engaged rather than top-down, listening rather than telling and participatory rather than science-led. Much of this book is concerned with providing examples of such interactions in practice. The aim of telling these stories is to offer lessons and learning points that can assist others seeking to carry out engaged water management. Interconnected with this change in the practice of water management is a second change, which involves a reorientation in relation to the nature of science itself. At the nub of this argument is the process of natural and engineering academics and professionals embracing an understanding derived from social sciences – that their science offers not a definitive truth, but one truth among many. In effect, embracing humility in relation to interaction with stakeholders also means humility in relation to science(s). Specifically, engineering and natural science professionals need to take on – and many already are - the idea that different valid truths are derived from a variety of methods, not all of which fit the model of 'good method' found in the physical sciences.

Before moving to elaborate further this argument about science, it is useful to clarify some terms and understanding about the 'engaged water management practice' that is the focus of this book. Here, the term 'participation' is used to imply processes through which the public have an input to defining or refining policy or practice. This might be at a national level but it might equally be at the neighbourhood level, such as in the example of the river restoration. In contrast, the term 'mobilisation' means that the stakeholders are involved in *doing* something that helps the policy to be implemented: for example, they might be saving water, harvesting rainwater or carefully selecting what substances are allowed into their drains. 'Engagement' is used as an overarching term to imply participation, mobilisation, or (and preferably) both. As will be discussed in later chapters, people are more willing to be mobilised if they have already participated and share with the water utility an understanding of water-related goals within their locality.

As the following chapters will argue, some key principles for achieving engaged water management are:

- engagement involves long-term relationships built on transparency in order to generate trust
- water utilities need to invest time to understand their stakeholders, not only as individuals, but also as members of formal and informal organisations and communities
- engagement requires that water utilities take time to explain their objectives to stakeholders
- being seen to listen and to understand community concerns is important; gifting resources to support communities is also useful, providing that it is not done in isolation when it can be read as buying a community
- engagement is most likely to be successful if it includes elements of participation and mobilisation together.

Finally, it should be noted that there is significant overlap between different forms of engagement and broader discussion of governance. As discussed in Chapter 3, governance is the broad process of overseeing a particular function for public purpose. Implicit in the use of the term 'governance' is the idea that other stakeholders, possibly other organisations or maybe members of the public themselves, have an input to designing and undertaking change. Processes of engagement – both mobilisation and participation – are therefore necessary if any governance functions are to be so shared. Because of this overlap there is a case for combining the language of 'governance' and 'engagement' and using just one term. This is not the route taken in this book. While I acknowledge significant overlap, in general I use the term 'engagement' when considering specific issues about how different groups interact, usually through purposive engagement processes. In contrast, the term 'governance' is used when the concern is broader, with how a particular watery function is overseen.

Beyond engagement: stepping outside the technocratic box

If many water professionals were to adopt an engaged approach to their planning and action it would form an excellent start to reconnecting people with water. However, there are three reasons why engagement alone is not sufficient.

First, effective engaged policy-making is not easy. Even as a committed water professional working in a supportive organisation, your scope for decision-making is not infinite: other agencies, regulators and landowners are also likely to have strong views. If you are lucky enough to be in a position from which you can bring them all to the table, there is still no guarantee that they will listen to each other or to others. It is also challenging and resource-intensive to bring the public (and we might ask which public?) to a discussion forum, and it is equally hard then to run a forum in which those who are not usually empowered have voice and input. Lessons from how policies have been made in other circumstances and, in particular, how and what has worked in terms of public engagement are all likely to be useful.

Second, there may be better ways to explore the views and behaviours of your stakeholders. You are more likely to get a fuller understanding of the thoughts and behaviours of different stakeholders if you (or a facilitator/researcher you employ) can find a way to interact with them in a relaxed environment where they can express their views without having to think about how they reconcile with others. If reconciliation is called for later, then, having built up trust and being seen to have listened, this would be a good starting point for examining how divergent views can be accommodated.

Third, to assume that answers to the challenges of achieving sustainable water management lie among your stakeholders (including the public) risks significantly narrowing your perspective. Stakeholders of the existing largely technocratic system may find it hard to imagine whether and how other systems could work. One means to help you and your stakeholders to think beyond the experience of the local water system is to draw on experiences from other locations and from other times.

These factors highlight the need for water professionals to widen their perspective to embrace a set of knowledge encompassing how policy decisions are really made, investigation of stakeholder perspectives and the nature of other water systems. But does this involve any more than adding 'social science' to the list of disciplines embraced as having relevant knowledge for the water industry? As I argue below, embracing the right sort of social science to carry out participatory and inclusionary research is a departure from the traditional way of doing science. Broadly, we can call this type of social science research 'interpretive'.

Introducing interpretive social science

Why do water professionals need interpretive social science?

Interpretive social science research emphasises the various understandings and meanings that different people have for a phenomenon, such as a water initiative. It is this emphasis on different meanings that makes interpretive social science of potential value to water professionals. For example, seeing the different ways that stakeholders understand a water initiative or water policy change can help to guide the water professional towards choosing the most effective route forward. Likewise, understanding how divergent views were and were not resolved in another governance situation can throw light on what is likely to happen. Through understanding others' perspectives, water professionals could be placed in a stronger position in which to reflect on the consequences of their activities, as well as to communicate more effectively with those whose views are different from their own. Interpretive social science therefore has the potential to fill the voids concerning values and politics that are left when contemporary water management seeks to move beyond the technocratic system.

The history of interpretive social science's development helps to explain how it has something different to offer water management. Interpretive research emerged in counter-definition to positivist research. Positivist social science draws on the methods and understandings of the natural sciences; hence, rigour is defined in statistical terms, and the objective is to generate models or predictions. This was the dominant approach to social science through the 1960s and 1970s (Connelly and Anderson, 2007). From the late 1970s onwards, however, social scientists began to react against this way of studying social phenomena. Positivist studies did not, they argued, do justice to the full range of social perspectives and phenomena observed. In particular, positivist social science made generalisations and drew inferences that obscured societal complexity, denying the insights and values arising from different cultures and traditions. Drawing variously on Marxism (Harvey, 1996), narrative traditions from the humanities (Berger, 1972), continental philosophers (Foucault, 1977), socio-cultural anthropology (Douglas, 1985) and feminist studies (Haraway, 1988), a 'cultural turn' sought to recognise the range and complexity of cultures and meanings, and hence to understand how these phenomena impacted on the physical and social worlds. In practice, after years of a schism between these different approaches to knowledge (Connelly and Anderson, 2007), contemporary social science demonstrates evidence of an emerging consensus in which both positivist and interpretive approaches are recognised as providing different but valuable insights (Sharp *et al.*, 2011b; Browne *et al.*, 2013a; Fam *et al.*, 2015).

Water engineers and scientists have already made significant use of positivist social science – for example, drawing on demographic forecasting in predicting future demand (Sim *et al.*, 2007). To date, interpretive social science has been drawn on less frequently within water management in the industrialised world. Limited previous use of interpretive research is perhaps not surprising; as interpretive social science breaks with the standards that technical training defines as 'good science', consequently, there is a potential for interpretive social science to be dismissed as irrelevant to the development of sustainable water management. It is therefore useful for water engineers and scientists to understand some of the basic tenets of interpretive social science so that they do not make hasty dismissals.

It should be stressed that the argument here is that interpretive social science offers a set of *complementary* understandings and skills that can help water managers understand and work within the worlds of public participation and policy-making. Interpretive social science cannot and should not 'replace' the important knowledge obtained from natural science or from quantitative social science that is already embedded in water management practice. Nevertheless, it is important for scientists and engineers to take on something of an interpretative understanding in order to view their science as one expertise among many (including both other research and the expertise of local people). It is therefore argued that an interpretive stance is useful for all science because it embraces greater humility.

How does interpretive social science differ from other research?

Like all science, interpretive research takes a critical stance towards taken for granted knowledge, and shares the belief that high-quality knowledge requires systematic and rigorous development. However, a key difference between the natural science tradition and interpretive social science is that interpretive scientists believe social researchers cannot take themselves outside of the social world in which they conduct research. This makes intuitive sense; unlike many purely physical phenomena (for example, water particles), people are likely to change their behaviour if they perceive that they are being examined. Moreover, while researchers' judgements about water particles' characteristics are unlikely to vary according to their social position, their views about other people cannot be formed in isolation from their pre-existing expectations and social understandings. For example, whatever I do, the fact that I am a middle-class white woman is likely to influence the way that Asian working-class men interact with me. Also, however hard I try, I probably cannot prevent my existing understandings about Asian working-class men influencing the way I behave in our interactions either. The logic is that complete neutrality in the study of social phenomena is not possible. 'Scientific' notions of rigour based on neutrality and objectivity therefore need to shift. Instead, in interpretive science rigour is created through transparency about the researchers' social and value position compared to those they are studying and the way that this changes through the research. By working with researchers' subjectivity, interpretive research allows the often suppressed issue of differing value positions to be brought into open discussion. Interpretive research therefore seeks to create rigorous knowledge while acknowledging the 'good sense' that social phenomena are inevitably 'interpreted'.

The consequence of this perspective is a number of differences between the way interpretive research 'looks' compared to research carried out in the natural scientific tradition.

1. *The primary goal is the production of narratives.* Whereas natural science and positivist social science focus on the characterisation of phenomena (often in the form of numeric or computer models) in order to enable prediction, the goal of interpretive social science is the production of narratives about how different value positions are played in processes of policy and practice (Connelly and Anderson, 2007). The narratives aid reflection and discussion about the impacts of policies and practices for those within the policy area under investigation, and for those who are regarding it from other fields. The descriptive 'stories' produced do not aim to tell the policy-maker what to do in a particular situation, but they do aim to help the policy-maker think through the situation they are dealing with. This emphasis on narrative or example can be seen to parallel the way engineering science draws on examples to understand phenomena.

2. *Generalisation is achieved through the creation of theory, but this may not be the key goal.* For interpretive researchers, individuals' specific local perceived context provides critical information for understanding phenomena. In order to properly understand context, interpretive social science emphasises depth not breadth – for example, case studies and interviews of a small number of people rather than a survey of a larger number. Generalisation is achieved through the development of social science theory, rather than through numeric generalisations. Theory offers a 'rule of thumb' through which insights can be crystallised, and quickly brought to bear on a new situation. Theory, for example, might involve a categorisation of people, institutions, or practices. Corroboration of theory occurs when it is applied and found useful in more and different cases (Sayer, 1992), and for Sayer (ibid.: 243) such corroboration is the key goal of interpretive research. For Flyvberg (2001), however, like parables and other stories in everyday life, case study narratives about policy and practice provide powerful examples through which readers will develop different understandings of their own situation and it is these that are the primary goal of interpretive research.
3. *To achieve transparency, researchers often explicitly write themselves into the narratives.* Researchers' 'interpretation' is understood to be present in their choices of whom to access and what to ask, as well as in the process of analysis (Yanow and Schwartz-Shea, 2006). Consequently, the researchers' changing understandings of the phenomena constitute an important source of learning, and the researcher needs to be explicitly 'reflexive' about their own subjectivity in the process of the research. This means that good interpretive narratives often include a discussion of 'I' or 'we' as the researchers describe how their perspective on a phenomenon changed during the process of the research. Rather than seeking to achieve the perceived impossibility of an 'objective' position, good research is characterised by transparency about the researchers' own understandings and how these developed during the research.
4. *Quality is recognised through strength of narrative, transparency, reflection and faithfulness to interpretive methods.* Finally, quality for interpretive research is measured through researchers' faithfulness to interpretive methodological and output expectations. For example, data analysis is carried out in a systematic manner through coding (classifying) sections of text according to themes, and through the explicit search for counter-instances, while interpretations are typically checked with research participants. The research output will be a convincing narrative about different perspectives on the phenomena evidenced with field data such as observations and quotations (ibid.). These measures of quality are quite different from adherence to scientific method as it is traditionally presented, which presupposes the possibility of achieving an external 'objective' perspective.

It is these differences that mean professionals schooled in technical approaches and natural science methods sometimes dismiss interpretive research. In particular, technical observers may be uncomfortable with the 'presence' of the interpretive researcher required for good interpretive accounts of research. Interpretive work may equally be dismissed because the depth of understanding achieved through a small number of interviews is not perceived as providing valid knowledge compared to a larger number of encounters, even if the latter are superficial. Finally, the case study approach may be perceived as offering limited broader generalisations, and hence there may be hesitation about according it the accolade of 'research'. As the above discussions have shown, these dismissals are based on misunderstandings about the nature and goals of interpretive research, and, I would argue, a misunderstanding about what can be potentially useful in informing water practice. The dismissal of such case study or 'anecdotal' evidence can also lead to very high expenditure on expensive surveys when better information could be gleaned from a more moderately priced qualitative study. If such dismissals are maintained then water professions risk missing out on insights about the differing values and societal choices about the world in which they are working.

Interpretive social science on water management

'Interpretive research' (ibid.) is an umbrella term covering a growing proportion of contemporary social science, including an expanding body of research on water policy and water management (Bakker, 2004; Shove, 2003; Southerton *et al.*, 2004; Strang, 2004; Sharp, 2006; Allon and Sofoulis 2006; Chappels and Medd, 2008; Taylor *et al.*, 2009). 'Post-positivist', 'constructivist', 'post-modern' and 'qualitative' are alternative descriptors that are sometimes used to refer to social science research conducted in the interpretive tradition. Social science research streams that investigate water management within the umbrella of interpretive approaches include 'political ecology', 'practices' and 'transitions' research. Additionally, historical and anthropological research on other cultures is often carried out from an interpretive perspective – though this is seldom stated explicitly. What makes such research interpretive is their shared perception that objective social observation cannot be achieved and, hence, their alternative emphasis on understanding what language and actions are taken by different groups, and how they are variously understood.

Water professionals are most likely to be familiar with interpretive research from qualitative case studies that usually (though not always) follow an interpretative approach (for example, Sofoulis *et al.*, 2005; Knamiller and Sharp, 2009). To date, interpretive researchers working in the water field have tended to pursue one of two strategies. Either they have done research on water but published their results within their own disciplinary field (for example, Bakker, 2004; Strang, 2004; Sharp, 2006, Allon and Sofoulis, 2006), or they have carried out hybrid research, which uses many of the insights and methods of interpretive approaches but mixes them with elements of positivist science. Important examples of the latter use qualitative interviews, but follow the positivist norm of assuming readers and the public share their strong value preferences (Olsson *et al.*, 2004; Brown and Clarke, 2007; Brown *et al.*, 2009). By explaining the nature of interpretive science to water professionals this book hopes to open up a third possible strategy – for interpretive research on water policy and practice to become publishable within water-related disciplines and journals.

One of the advantages of the interpretive approach over conventional positivist social science relates to its relevance to the contemporary challenges faced by water professionals. Conventional positivist social science (for example, a survey) seeks to understand people studied through pre-existing and defined understandings. It assumes that the researchers know enough about the views of the people they are researching to define the most important and relevant questions. Sometimes this is a valid assumption but often it is not. In contrast, the conversational and observational approach of interpretive research provides space for the people being researched to express their perspectives about all aspects of any 'improvement' proposed. While the results of interpretive research may be drawn from a narrower context and fewer people than in a survey, the results at least reveal, rather than hide, the range and complexity of people's responses to the changes proposed. Such results enable water professionals to anticipate future potential problems. They are also often cheaper than commissioning a large-scale survey. However, as will be discussed subsequently in Chapter 5, quantitative approaches are increasingly being used alongside interpretive work to reveal new combined insights.

Conclusion

This chapter has introduced the idea of a 'technocratic model' of water management in which water is seen as a resource to be exploited for the provision of water services to ever more demanding agriculture, industry and citizens. Any difficulties with this model have been assumed to be surmountable through better science.

As the chapter has narrated, the technocratic model has come under increasing question in recent years. Attempts to develop technocratic approaches through the concepts of integrated water resource management, ES or AM have moved some way to rhetorically questioning the assumptions of previous water management. No longer are the public assumed to be completely passive and uninterested in water decision-making. Likewise, the physical environment is increasingly recognised as having a valid need for water. Similarly, within many pockets of practice, the public is increasingly involved in decision-making and action.

Despite these significant gains, there remains a naivety in the implicit understanding of the public within these models and practices. Given the technical training and background of many of those administering water-related public participation, they have made impressive efforts to open their practice up to scrutiny. But combining a variety of scientific and lay perspectives is not easy, and it does not automatically lead to an agreed route forward. In particular, the challenges of dealing with a plurality of different values are issues that water managers should not be struggling with alone. In other fields – most notably the humanities, but also in practical fields like town planning and social work – research fields and methods have been developing to study and address the difficulties faced in fields where values compete. The resulting research – here termed 'interpretive research' – provides a rich source of methods and approaches that can be of assistance in relation to water management. By drawing on and accepting interpretive science as valid science, water managers can learn from many 'stories' of how others have managed water and water engagement, and this can be a route to considering and developing their own practice.

Overall then, this chapter has made contributions to all four arguments set out in the Introduction. By introducing and describing 'technocratic water management', this chapter has laid the framework for both the historical and contemporary 'descriptive' arguments. Technocratic water management, it is suggested, is the means through which 'disconnecting' occurred, cutting individuals off from the peculiarities of their locality, and offering the promise of a universal and one-size-fits-all service model. Likewise, contemporary attempts to develop the technocratic model by reconnecting with stakeholders show signs that some changes in water management are underway.

In terms of a contribution to the normative argument – that more and better 'reconnecting' is needed – a case was made that it is challenging to agree routes forward when there are diverse perspectives among stakeholders. It is important that the complexity of such exercises is recognised. Again, what is presented here is the key framework of the argument, but subsequent chapters are needed to provide illustration and nuanced understanding of what this complexity means.

The latter part of this chapter introduced readers to interpretive research, making the case for the need for a new form of science more able to accept a plurality of understanding and perspectives. The discussion was theoretical – it suggested that, contrary to the intuitive understandings of many schooled in the natural sciences, interpretive research can offer good science with useful and usable insights. This claim will be illustrated through interpretive research cases and theory in subsequent chapters.

2

URBAN WATER USE IN CONTEXT

This chapter tells a physical story about total global quantities of water and where they go. In this respect, it puts the 'urban water management' discussed in the rest of the book into the broader context of global water resources. It also takes a chronological journey through water resource management concepts, and hence through changing concerns about water resources. Specifically, it explores the water cycle, 'blue' and 'green' water, 'consumptive water use', water footprints, virtual water and new attempts to conceptualise a 'hydrosocial cycle'. As such, it is one angle on the rise and putative critique of technocracy.

At the centre of this chapter lie the positivist concepts through which global water resources are described, modelled and enumerated. By demonstrating the limitations of these quantitative understandings, the chapter also lays the foundation for why a more connected and qualitative interpretive form of science is needed.

The sources supporting this chapter are a mixture of academic works, effectively drawn from two broad areas of scholarship. On the one hand, those producing most of the concepts described are hydrologists and water resources experts, many of whom are skilled in technical understandings and statistical manipulation. On the other hand, Jamie Linton has led academic geographers to assess and critique these quantitative understandings.

The water cycle

The water cycle is perhaps the most well-known scientific understanding linked with water. Figure 2.1, shows the processes of precipitation, surface run-off and infiltration, as well as transpiration and evaporation. Diagrams like Figure 2.1 are very familiar to all of us – they are taught in schools and colleges as a means of explaining the links between different 'natural' water processes.

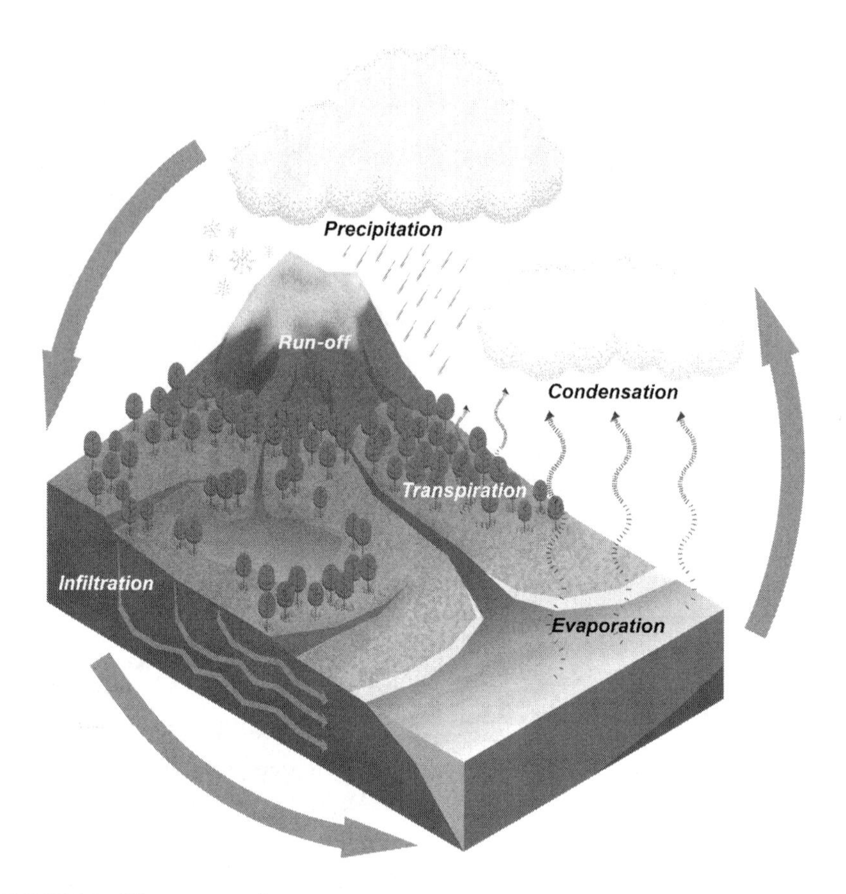

FIGURE 2.1 The water cycle

While some understanding of the scientific processes that we now see as part of the water cycle go back centuries, Linton (2008) traces the development of a cyclical understanding to the early 1930s. Linton points to a schematic by US hydro-geologist Horton in 1931 as crucial in bringing water processes into the realm (and intellectual ownership) of hydro-geologists. However, it was the US National Resources Board in 1934 which first presented a 'naturalistic' diagram depicting land, air and sea, which is instantly recognisable as the water cycle, and similar to that shown in Figure 2.1. Behind the apparently simple diagrams lies a series of calculations about quantities of water in any part of the cycle at one time, together presenting water as part of a 'dynamic and self-contained mechanism' (Linton, 2008: 636), based on the equation that available water is equal to precipitation less evaporation. In depicting the cycle, Linton suggests that state agencies sought to make the water cycle more visible, and hence to argue for better monitoring about and more exploitation of water across the whole of the USA.

For Linton, the classic water cycle's depiction of 'natural' processes only maintains a nonsensical idea, of the human imagination, that humans are separate – and separable – from nature. It is ironic, then, that the very depiction and discussion of the water cycle was part of the process of humans intervening on a much more systematic scale in water management across the USA from the late 1930s onwards, and subsequently in other countries. Specifically, the water cycle was used as the justification to call for monitoring rainfall and river levels, knowledge that then enabled proposals for water resource 'development' to be planned and acted upon (Linton, 2008). Though the drawing depicts only natural processes, its anthropocentric perspective is betrayed by some of the language with which it is associated. Evaporation and transpiration are typically depicted as 'losses' (and, indeed, this language dates back to Horton's perspective in 1931) (Linton, 2008: 641). Similarly, the water behind dams and in streams and aquifers is often referred to as 'renewable water resources' (FAO, 2014), or 'available water' (Green Facts, 2014), implying its usefulness to humans. In practice, if all such 'available water' were taken from a river the ecosystem would be ruined. (As a consequence, I will discard 'available water' and use the more recent term 'blue water' to describe water resources in 'blue' rivers, lakes and aquifers from here onwards.)

A second issue raised by Linton relates to geography: he argues that the hydrological cycle 'irons-in' a particular hydrology that relates to the water cycle's origin on the east coast of the USA (2008). Specifically he suggests that the 'even' treatment of evaporation, condensation, precipitation and run-off is inappropriate to other climatic conditions in which particular sets of these processes (e.g. evaporation and condensation in deserts) are massively dominant. Likewise, the cyclical depiction of the water 'cycle', combined with the trees and valleys frequently pictured, imply some stability of climate from year to year in a way that does not work for many arid or tropical climates in which annual precipitation quantities are highly variable between years, and within the year. Finally, the focus on available water denies the importance of precipitation and soil moisture in providing 'natural' irrigation to ecology and to rain-fed agriculture. In all of these respects, the water cycle is more or less appropriate and useful at its origin, but becomes increasingly inappropriate if it, and related concepts, become universalised for discussion of water resources everywhere.

Before moving on from the water cycle it is useful to reflect on what it says about 'urban' water use. The key point here is that the water cycle is primarily concerned with quantities of water. And from this perspective the urban use of water is not important at all – we may withdraw water for use, but almost all of it is then returned to the streams or other bodies of 'available' water after use and treatment. In this respect, urban water use is 'non-consumptive' because what is withdrawn is later returned. With this focus on available 'blue' water resources, then, and with minimal consideration of water quality, urban water use is, indeed, dismissible.

Water scarcity

The development of the water cycle in the 1930s was associated with a concern with national water resources, and with ensuring that such resources could be effectively utilised so that the country could develop. In this and the subsequent section, I describe a shift associated with the 1990s and 2000s in which the focus moves on from national resources to total global resources, linked to wider concerns with population growth, and future food and energy supplies. As I will argue, some adjustments are made from the language of the water cycle to take account of contemporary concerns, but other lacunae associated with the water cycle continue.

Linton (2004) highlights a significant change in the way that water resources were discussed that emerged in the 1990s. Whereas previously the concern to quantify water resources related to the catchment (whether in a national or international setting), the new focus was on global water resources, with boundaries usually drawn at a national level. A key factor enabling these studies to be carried out was the emergence from the former Soviet Union of techniques for estimating water resources in regions where there was limited data. Although the calculations were extremely rough estimates, they enabled a statistical field-day; quantifying and mapping global water resources alongside numerous other global statistics generated a number of new concepts and indicators.

Primary among the new concepts to emerge was what has become known as the Falkenmark water stress indicator, a measure of the national water availability per capita. The Falkenmark indicator is claimed to relate to the quantity of water needed for individual usage; in other words, 'absolute water scarcity' is indicated when the water available is less than that which is (judged as) required to serve potable water, washing and food growing needs (Falkenmark, 1989; Brown and Matlock, 2011). Table 2.1 shows the trigger levels at which water scarcity is perceived to give way to water stress.

In reality, as Gleick (1996) highlights, the domestic or 'urban' component of this demand is minimal. Per person daily water requirements for drinking (5 litres), sanitation (20 litres), bathing (15 litres) and food preparation (10 litres) together total 50 litres, a tiny proportion of the 1,200 litres per day that Gleick sees as the total daily requirement. While it is possible to criticise such estimates as universalising, and as not recognising the variety of cultures and their differing expectations of water, these estimates are useful in putting urban and domestic uses into the context of proportionately much larger agricultural water demand.

TABLE 2.1 The Falkenmark water stress indicator

Level	*Quantity of available water per person per year (m^3/cap/yr)*	*Equivalent quantity in terms of litres per person per day (approx.) (l/cap/day)*	*Examples of countries (UN Water, 2013)*
Absolute water scarcity	<500	<1,200	Libya, Saudi Arabia
Water scarce	500–1,000	1,200–2,400	Morocco, Egypt
Water stress	1,000–1,700	2,400–4,660	Denmark, Poland
No stress	>1,700	>4,660	Portugal, UK, Australia

The Falkenmark indicator suggests that water stress is focused in North Africa and the Middle East, but that there is no water stress in the USA or Australia. This betrays some of its failings as an indicator: national level averages hide large differences within a country, and the allocation of one quantity of water which is needed per capita for human survival (including arable agriculture) clearly ignores huge variations in climate, culture and agricultural practices between countries.

One development of the Falkenmark indicator sought to factor in human and environmental uses, hence seeking to make good some perceived failings of the water cycle (Comprehensive Assessment of Water Management in Agriculture, 2007). The International Water Management Institute therefore distinguishes between 'physical water scarcity' and 'economic water scarcity'. Physical water scarcity occurs when there is not enough water to meet demands (including estimated demands of the environment), either because of low precipitation, or possibly because of large demands on the water resources from agriculture. An example of a region classified as physically water scarce is the south west of the USA. In contrast, economic water scarcity occurs because there is not sufficient investment in water resource infrastructure or in effective institutional management such that the available water resources are fully used. Much of Africa south of the Sahara is so classified: the resources are there, it is implied, but more investment is needed to ensure that they are effectively used.

The concepts of physical and economic water scarcity provide broad quantitative classifications of regions that may 'make sense' in relation to our intuitive understandings of what is needed: regions with economic water scarcity need investment to make use of the water they have, while those with physical scarcity need to meet new water demands by reducing the quantity needed for another water demand. Very roughly, they split into more industrialised countries that are short of water and emerging economies that are short of water.

However, like the water cycle, the focus on water scarcity was still concerned only with the visible 'blue' water resources.

Blue and green water

In many localities it is rain and local soil water that irrigate food production, not the water 'available' in rivers and behind dams (Oki and Kanae, 2006; Falkenmark, 2008). If we are to discuss total water resources, then this is a serious omission. Indeed, Mekonnon and Hoekstra (2011) calculate that such unseen 'green' water makes up 74 per cent of the total water used by humans.

Shifting from 'blue' to 'green and blue' water also means changing the focus of measurement: rather than considering 'stocks' of water, the new focus needs to be on flows. A blue-green focus then asks from which sources (blue or green) and through which routes (the landscape, rain-fed agriculture, irrigated agriculture, or industrial use) is the water being used?

Focusing on the combined use of water by humans, Mekonnen and Hoekstra (ibid.) estimate that 92 per cent of water is used for agriculture, 5 per cent for industrial usage and 4 per cent for municipal and domestic use. The key gain with this way of calculating is the benefit of rain-fed agriculture – for example, most of the agriculture in the UK is recognised rather than dismissed. The quantifications of blue and green water have allowed new comparisons between places to be made based on how water is embedded in agricultural and industrial use. A further comparative vocabulary has blossomed based on the trade in 'virtual water' (Hoekstra, 2003) and different locations' 'water footprints' (Hoekstra and Chapagain, 2007).

Mekonnen and Hoekstra (2011) have compiled accounts of the total annual water footprint of consumers in different countries – that is, the total amount of blue and green water consumed per capita. Relating both to the level of development, and the extent of meat in diets, national per capita water footprints vary around an average of 1,385m^3 per year. The water footprint of some countries is shown in Table 2.2.

The right-hand column in Table 2.2 shows the water footprint in terms of litres per person per day. This can be compared with municipal water use per day that is taken for the public water supply – which is about 150 litres per person per day in the UK (Defra, 2008). This again highlights how, in water footprint terms, domestic and municipal use is of minimal importance compared to food and other goods.

Given the high proportion of water resources that are used in agriculture, it is not surprising that most water footprints are made up of 'virtual water' – that is, water that is consumed through food and products purchased from other nations. Some countries (e.g. the UK, Japan and the Netherlands) are importers of virtual water, others are exporters (e.g. Australia, Argentina and the USA) and yet others are both importers and exporters (e.g. France and Germany) (Mekonnen and Hoekstra, 2011: 21). However, such trade in virtual water is not necessarily a bad thing. Given the weight of water and the energy and logistical challenges of transporting it, it is more environmentally and financially sensible that such goods should be traded rather than the countries exchanging the water itself and then growing/producing their own goods (Allan, 2003). Virtual water in the form of imported food can support habitation in arid cities such as Phoenix, Arizona.

TABLE 2.2 Water footprint in selected countries

Country	*National average water footprint (m^3/cap/year)*	*Equivalent in litres per day (l/cap/day)*
Global average	1,385	1,385*1,000/365 = 3,795
Bolivia	3,468	9,501
USA	2,842	7,786
UK	1,258	3,447
China	1,071	2,934
India	1,089	2,984
Democratic Republic of Congo	552	1,512

Source: Mekonnen and Hoekstra, 2011

One of the important benefits of talking about virtual water is that we can start to explain some areas of water shortage. Much of Australia is short of water not because of low rainfall (or certainly not low per capita rainfall) but because of agricultural uses of water that are very demanding of these resources. Likewise, one political impact of water footprints is to highlight the UK's complacency about water security. There may be sufficient rainfall in the UK to feed much of our own agriculture, but that is only because we import so much virtual water from elsewhere (Norton, 2014).

Calculations about water footprints and virtual water have been important inputs to global debates about the so-called 'water–food–energy security nexus' (e.g. at the Bonn nexus conference 2011, see Stockholm Environment Institute, 2011). These discussions highlight some of the fundamental choices that must be made in terms of how global water resources are directed: is water used for growing food, for generating hydro-power and/or for growing bio-fuels? How can global water resources be managed to cater for the combined demands of a growing and developing population for energy, food and water? These questions are addressed through a recent out-pouring of works on the 'nexus' (e.g. Allan, 2003; Hoff, 2011; Cairns and Kczywoszynska, 2016).

An issue with the focus on water footprints and virtual water trade is that the data say nothing about the scarcity – or otherwise – of water in the location where the goods were made. Tomatoes from different countries are evaluated in a virtual water calculation according to the amount of water taken to grow and package them. However, 10 litres of water growing tomatoes in Israel, where water is very short, should probably be seen as more 'expensive' in terms of the water's relative value (and hence, socially and environmentally problematic) than 10 litres of water from Italy, where water is more plentiful. Virtual water and water footprints are useful political tools to help us to think more about water and its value; but, as with much quantitative data, it lacks specific information about who is gaining or losing from water allocations in specific localities.

Further developments of the virtual water and water footprint thinking have sought to overcome perceived problems with these water resource management concepts. In his 2008 review of sustainable development and water management, Falkenmark stresses the important difficulties that result from the pollution of waters. In their 2011 work, Mekonnen and Hoekstra go beyond just green and blue water to also consider what they call 'grey' water, and through this to address the lack of attention to water quality in previous calculations. Mekonnen and Hoekstra use the term 'grey water' to refer to the amount of (blue) water needed to assimilate pollution based on existing ambient water quality standards – in other words, to dilute the polluted water to 'normal' in the context of their surroundings. Whereas green and blue water normally flow physically out of the system, this thinking sees 'grey' water lost to the stock of water resources because it must dilute the pollution. While some attention to water resource quality is welcome, this understanding can be slightly confusing; whereas the words 'blue' and 'green' indicate the source of the water, the term 'grey' is an indication of how it is used. It should also be noted that this meaning for the term 'grey water' differs from that utilised in urban water discussions (when grey water is untreated or partly treated used water that is available for reuse).

Hoekstra elaborates the footprint concept in a 2014 article in which he proposes three new indicators that could be used as a basis for water allocation (Hoekstra, 2014). First, he suggests that water footprint caps per basin could ensure that water use within any one basin did not exceed that which its ecology could sustain. Second, he indicates that water footprint benchmarks per product could support a new process of valuing (and selecting) products based on their water footprint. Finally, he suggests that a 'fair water footprint share per community' could contribute to a debate about whether and how water should be equitably distributed. The cleverness of Hoekstra's indicators is the suggestion of the use of the three indicators together. Limits on water use per basin are clearly appropriate and useful in limiting local use of a local resource (including local use for export), but reveal nothing about how the same community is consuming virtual water. Footprints per product and ideas about 'fair water shares' have the potential to start to make us think about the impacts of our water use elsewhere. Only when used together could the indicators start to ensure that water is distributed in a less environmentally damaging and more socially equitable way than it is today; tellingly, if used together, these measures could address the Israeli tomato problem highlighted above.

Hoekstra's development of the water resources quantification process points out that the allocation of water is essentially political. Nevertheless, the practical use of his indicators to guide water resource allocation is a provocative fantasy; it could occur only in the context of an environmentally benign world dictatorship. It remains useful, however, because it helps to highlight how financial resources command water footprints and environmental damage.

The urban water cycle

Understandings about consumptive and non-consumptive water resource use, the impacts of land use and the pollution of water resources have been translated into the popular imagination through the depiction of an 'urban water cycle'. As in the example in Figure 2.2, the term 'urban' is a misnomer, as 'rural' processes of the capture of water and its use in agriculture are shown as well as water uses in the city. The difference between this and the conventional water cycle is that human actions are factored in, including impacts on land use and flows, pollution as well as water use. Nevertheless, we can critique the cycle as depicting only the 'direct' actions on water or on the land immediately surrounding it. The urban water cycle does not depict 'indirect' actions, such as changing the price of water, providing flood warnings, or permitting households to make their garden impermeable.

The urban water cycle is also approached academically, for the purposes of modelling and enumeration of flows. In a paper reviewing how the urban water cycle is modelled, Urich and Rauch (2014) provide a general depiction of how models portray the urban system, shown in Figure 2.3. Their focus is on how cities adapt to

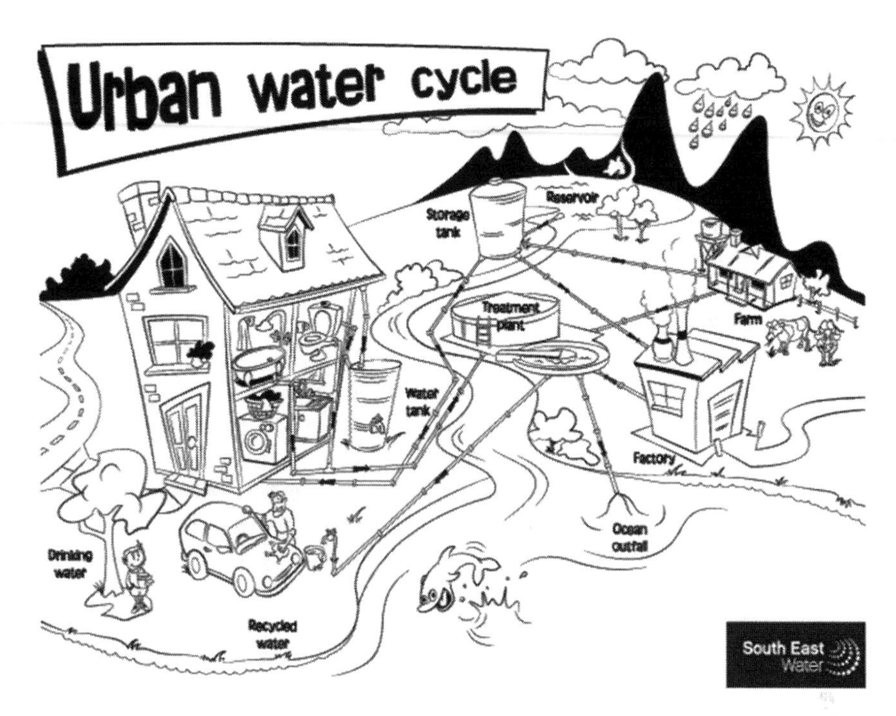

FIGURE 2.2 A children's colouring example of the 'urban water cycle'
Source: South East Water, 2016

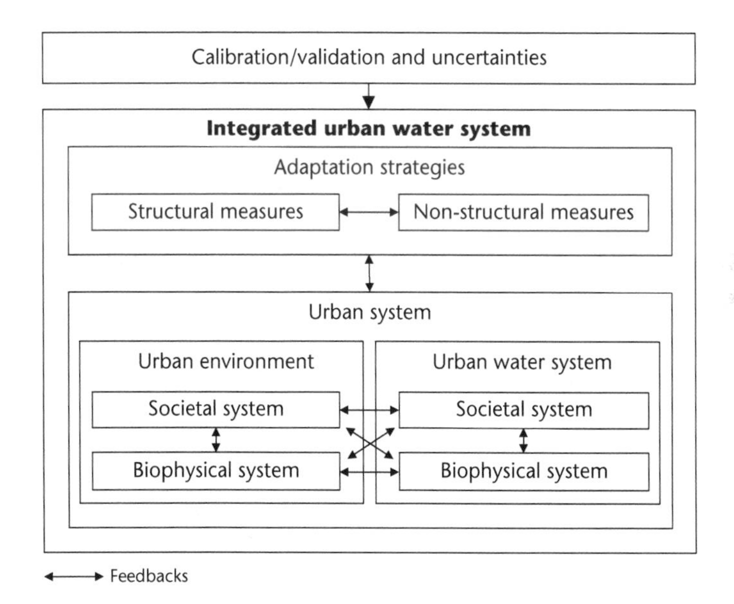

FIGURE 2.3 Dimensions and linkages of modelling the urban water system
Source: Urich and Rauch, 2014, with permission from the copyright holders, IWA Publishing

climate change, distinguishing between structural measures (e.g. engineered water storage tanks) and 'non-structural measures' (e.g. landscape features or the addition of water tanks to household water supply). They separate the urban environment from the urban water system, and within each they distinguish the biophysical flows and the social system. From this depiction we can see a concern with the different practicalities associated with modelling the familiar structural measures from the newer non-structural measures, where there are greater uncertainties. There is recognition that the water system itself influences not only the way water flows, but also the environment in which it is set. Notably, in their depiction of modelling processes, the 'social system' interacts with the water and non-water environment. In the paper, they describe how most models of the urban system explore the environment's evolution through time using 'agent-based models' in which geographically focused demographic statistics provide indications about how populations will react to choices, for example, about the source of their water supply. Nevertheless, they stress that these processes of modelling how social factors impact on the environment are new and difficult. For example, they highlight how measures exerting indirect effects on the water system such as insurance are rarely modelled.

Compared to the conceptualisation of water resources, both the popular and academic models of the urban water cycle give a higher emphasis to land use and water quality. In this respect they acknowledge a wide range of human impacts on the water cycle. However, society is black-boxed in a one-dimensional way, focused on the direct influences on the water system or its environment. Warnings, charges, media reports and other indirect influences or 'intra-societal interactions' are not considered in these conceptualisations.

The hydrosocial cycle

The various water indicators discussed above might all be seen as quantifications of and elaborations about the classical water cycle: elaborating the underground processes through discussion of green water, and factoring humans in through considerations of water allocations per capita, and embedded water in products. Linton and Budds' 2014 paper argues that the hydrological cycle needs to be replaced because it is 'a social construct with political consequences'; specifically it builds on the a priori division of humans from water and it has allowed the exploitation of water resources for profit. In their view, changing historical circumstances now 'favour the introduction of new ways of conceptualising water so as to reflect and draw attention to its social dimensions' (ibid.: 172). Consequently, they present the 'hydrosocial cycle', a concept that has been developed and promoted by a series of geographers working in the political ecology tradition (Linton, 2008; Swyngedouw, 2009; Linton and Budds, 2014).

Linton and Budds' depiction of the hydrosocial cycle is shown in Figure 2.4. It is much simpler and less figurative than the hydrological cycle it seeks to replace. For Linton and Budds, 'water' (in the centre) represents the different types of water that are produced by different hydrosocial configurations. In contrast, H_2O represents

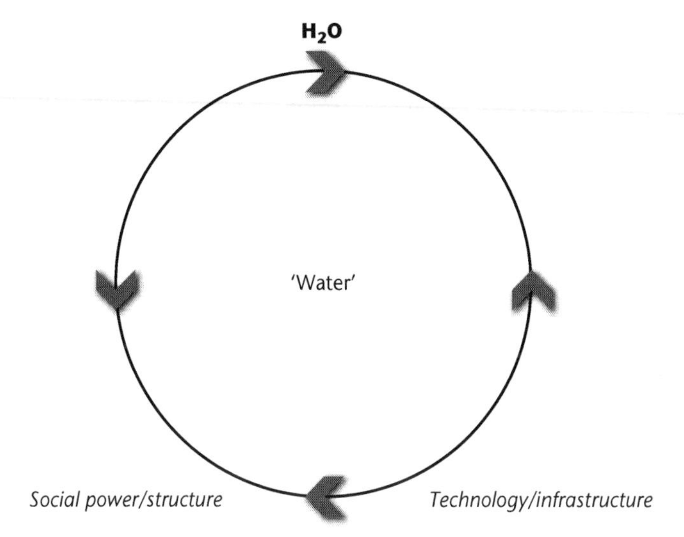

FIGURE 2.4 The hydrosocial cycle
Source: Redrawn from Linton and Budds, 2014: 176

the materiality of water and its ability to stabilise/disrupt. The two-way flows of the diagram highlight how society's impact on water (and vice versa) can be direct or may be mediated through technology and infrastructure.

The central argument associated with the hydrosocial cycle is that 'particular types of social relations produce particular types of water, and vice versa' (Linton and Budds, 2014: 175). For its advocates, the hydrosocial cycle also highlights how the material nature of water (depicted as H_2O in Figure 2.4) can have an active role, sometimes structuring social relations, but, equally, sometimes destroying them – for example, through a flood or a drought. The hydrosocial cycle therefore calls for 'revisiting traditional fragmented and interdisciplinary approaches to the study of water by insisting on the inseparability of the social and the physical in the production of particular hydrosocial configurations' (Swyngedouw, 2009: 56).

Linton and Budds emphasise the analytical advantages of the hydrosocial cycle. They argue that it is a means to explore how different specific waters are produced and are valued differently by stakeholders, as a guide for uncovering the power relations in water resource exploitation and distribution processes and as a means to highlight the importance of the different ways that water is known (Linton and Budds, 2014).

Like the urban water cycles depicted in Figures 2.2 and 2.3, Linton and Budd's understanding of their hydrosocial cycle emphasises the role of technology and infrastructure in influencing and being influenced by water. In its discussion of social power and structure, the language associated with the hydrosocial cycle goes beyond the direct human influences on the water cycle. Nevertheless, the two-way circulating arrows make this a confusing diagram to follow. Apart from indicating that waters, social power and technology are mutually influencing, it is hard to see what else is communicated. In particular, the depiction gives little guidance about how processes of water management and governance occur.

A hydrosocial governance cycle

This book shares significant interests with the political ecologists writing about the hydrosocial cycle. However, while, through the cycle's focus on political ecology, it takes an explicitly critical stance and seeks to highlight inequalities, here the aim is broader. The question is not 'how can we put people into our analysis of water?' but rather the subsequent one, 'how can we put people into management of water?'. In light of this more practical focus on the management of water, is there a more useful way of depicting hydrosocial interactions than those shown in Figures 2.1–2.4?

Figure 2.5 shows a hydrosocial governance cycle. Like the hydrosocial cycle, there is a concern to depict 'waters' rather than water, and these are shown sitting within a physical environment. This environment includes both 'natural' and engineered features, with no separation between these. The point is that the physical world influences the water and vice versa, but that seeking to separate which parts of that physical world are natural and which are human-engineered is a distraction. People are described as 'individuals and communities' because they may be acting as individuals, as households, or in larger groups (communities). People sit separately from another social category that I have called governance. Governance (discussed in more detail in the next chapter) can be seen as processes of 'steering' of collective systems – in this case steering of 'water management', which can occur through concepts or through organisations. For the purpose of the diagram, the informal interconnections that make up communities sit separately from formal 'organisations' shown within the Governance box (in practice, it is acknowledged that the line between the two can be fuzzy as loosely connected communities can cohere and become organisations). Organisations such as water utilities exercise governance through a number of routes. They can influence waters directly (e.g. through treating

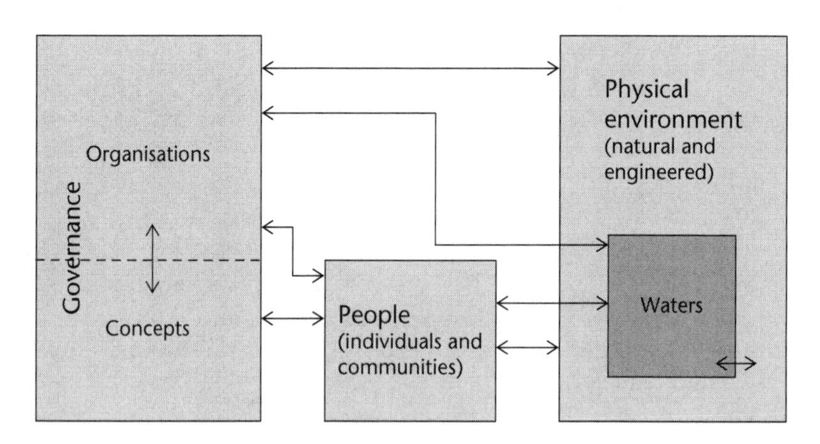

FIGURE 2.5 A hydrosocial governance cycle

them), but they might also influence the physical environment around the water (e.g. through controlling land use), or they might influence the way that people are using water (e.g. through new regulations about what can be put down a drain). For example, pre-1800 sanitation had to be regulated stringently because it was managed by households, whereas today far less stringent regulation of households is possible because waste disposal is managed within centralised and heavily integrated systems which require relatively minimal householder compliance (Skelton, 2016). Although the physical environment has influenced organisations (e.g. many water utilities are formed within drainage basins), this occurs less often than organisations influencing waters. In relation to concepts, a particular depiction of the world (e.g. the water cycle or water footprinting) effectively suggests factors that should influence the rearrangement of waters. Such concepts exercise influence via organisations, or otherwise through people.

Figure 2.5 has no quantifiable circulation of water, as occurs in the traditional water cycle. In this strict sense, the word 'cycle' should no longer be applied. In a looser sense, however, cycle may be an appropriate word. The processes depicted are dynamic and the nature of all 'boxes' is in a constant state of flux, driven partly by the interactions shown. Various cycles (e.g. the physical water cycle, policy review cycles (Hill, 2013) and the 'issue attention cycle' (Downs, 1972)) occur within the boxes shown.

For those who think in graphical terms, Figure 2.5 may summarise some of what this book is trying to discuss. As has already been argued, technocratic water governance focuses on the ways that organisations were able to directly influence waters – depicted by the arrows at the top of the diagram. In contrast, hydrosocial water management emphasises that water governance involves both direct influences on the physical environment and governance influences via 'people'. In this respect, hydrosocial water management can be seen as involving all of the arrows shown.

Diagrams do not show complexity or richness of understanding. In this respect, Figure 2.5 should not be expected to convey the breadth of material covered in this book. Nevertheless, the figure is useful as a schematic representation of how the key elements associated with water management are understood.

Conclusion

In terms of the hydrosocial governance cycle shown in Figure 2.5, this chapter has focused on the 'concepts' part of the Governance box. Specifically, it has been concerned with changing ideas about water supply. These are important partly because they influence governance practices; for example, Linton argues strongly that the hydrological cycle has ironed in an understanding of water as an exploitable economic resource. Even when concepts are less influential than the hydrological cycle, the metaphors they introduce (e.g. the water footprint) highlight different water problems, reflecting the current concerns of water resource managers, and in turn likely to have some influence on water management decisions. Certainly, the various water cycles, water scarcity and concepts of virtual water and water footprints are part of the vocabulary through which we think about and discuss water at both a local and a global scale. While many of the concepts frame their understandings as neutral technical observations, some (most notably Hoekstra's three indicators) are explicitly political, using their science to call for more careful and equitable distribution of resources.

Conceptualising, modelling and representing the world is an important role for science as part of governance. Nevertheless, a key message of this chapter is that all such concepts should be used with care. The key reason that the concepts should be treated with care is because they are often discussed as universal concepts when, in fact, they unhelpfully generalise particular local experiences. Each selection of specific metrics to be compared and mapped imposes a particular political understanding about the important differences between localities. In this respect the concepts are an instrument of power through which various (both less and more progressive) western understandings about water exploitation are elucidated and imposed on others.

To understand the issue of universalisation, it is perhaps useful to imagine yourself as a teacher attempting to educate rural children in North Africa about the classical water cycle. In farms which are dependent on rain-fed agriculture, what is the relevance of 'available water'? Trees depicted growing naturally on a hillside may not coincide with a child's experiences where trees are carefully nurtured at locations where occasional storm rainfall can be stopped and infiltrated. In this context, infiltration of water from a temporary stream during storm is not a 'loss' to the resources, but a crucial gain, to be maximised through walls and other devices within the dry riverbed (Molyneux-Hodgson, 2015). Of course, you can (and should) develop a variation of the water cycle that picks up and depicts these local experiences so that the concept does make sense to the children. But the fact that you have to do this, and that their experience is not depicted or communicated otherwise in this thing called 'science' in the textbook or on the web, is important. The lesson for western scientists – not just those of water resources, but all of us dealing with knowledge – is to be aware and sensitive to the particularities of our knowledge and conceptual understandings, and hence to be curious about and sensitive to how and why they don't work, and to highlight and develop variations in order that a wider variety of experiences are depicted and explored.

In this context, some particularities of the technocratic water resources perspective and concepts of which we need to be aware and for which we may need to compensate include:

- a universalising focus on water as one substance of equivalent quality across space and time, where the main issue of importance is the overall water quantity
- a focus on annual average quantities of water as the critical factor of concern, with variations in the pattern of arrival within the year, or between decades, not recorded or seen as relevant
- a concern with how water can be best used to aid human development, largely focused on commercial food growth and hydrological or bio-energy production, with limited emphasis on subsistence agriculture

- a universalising understanding of human needs, formed around a relatively austere version of western lifestyle, diet and energy demand, but no recognition of variations between cultures, or of human ingenuity in the face of shortage
- alongside the minimal emphasis on water quality (or qualities – for these vary widely), the dismissal of urban water use as irrelevant as it is non-consumptive
- a black-boxed understanding of intra-societal processes, which are seldom considered or depicted.

In terms of this book's focus on reconnecting people with water in urban areas of the industrialised world, this chapter makes three substantive contributions. First, in chronicling the development of water resource concepts, it provides one angle on the nature of the technocratic water management regime and its internal machinations. As such, it contributes to 'argument 1', as presented in the Introduction, because it describes how water resource management concepts promote a water management that is largely devoid of people. Second, it has pointed to the need for the addition of both more and different quantitative measures (some of which have been discussed), but also for qualitative understandings, informed by less universalising approaches such as interpretive social science and the hydrosocial water governance cycle. Thus it contributes to the fourth argument discussed in the Introduction, that more interpretive science is needed. Third, the chapter highlights how in the context of water resource management, urban water and its impact on water quality is dismissed as relatively irrelevant. In this respect it poses a challenge, asking why, in the context of a more 'hydrosocial' perspective, is it suddenly important to focus on reconnecting people with water in cities?

There are several answers to this question. First, and most obviously, the world is urbanising and the majority of people now live in cities. Second, and perhaps more importantly, the water resource measures discussed in this chapter may feel irrelevant to most readers; unless you are a farmer or a water resource manager these discussions are outside your realm of knowledge. Yes, the process of reconnecting people with water may well have to take on and discuss the water embedded in food and other goods, but at present even the water coming out of the tap feels distant from most people's understandings. If we want to move towards a world in which politically sensitive measures of water management – such as Hoekstra's trio of indicators – are used, we must first find ways to help people to connect to, to understand and to value water in their home and their local environment. Third, and closely linked to the second point, we have already seen that these quantitative measures give little importance to the qualities of water flows, and yet these are the factors which are most impacted on by cities, causing damage to ecosystems but also to human livelihoods. So, understanding and protecting water quality creates a greater focus on the urban, and on urban experiences of water management. Finally, just as water quality is a detail that goes beyond the universalising quantification carried out in water resource calculations, so issues about the rhythms of water flows are hidden from view. Again, we know that cities create profound changes in water infiltration and flow patterns, often exaggerating natural extremes, and making floods and droughts more severe.

Overall, this chapter suggests that a hydrosocial approach should take cognisance of and utilise a variety of quantitative measures of water resources as part of forming and debating political understandings about water resource distribution. However, it also stresses that such measures are not enough, and that to truly reconnect with water there is a need to focus more on the background and experiences that most people have of water, and that means focusing on the urban context and thinking about the management of water quality and of flood risk management as well water supply.

3

THE GOVERNANCE OF WATER SUPPLY AND DEMAND

This chapter explores changes in the way we have thought about and managed water supply. From a contemporary perspective, provision of a piped urban water supply is a crucial marker of development and modernisation. But it was not always thus. Moreover, contemporary over-exploitation of water resources is calling into question the assumption that water is abundant. In examining the changing thinking about water supply and demand, therefore, this chapter provides another part of the story of the rise of and challenges to technocracy.

Water supply can be seen as one of the most immediate and obvious 'collective goods'. It quite literally saves effort and money if a group of neighbours work together to ensure a clean water supply, rather than each household looking after their own. However, if we are going to work together to share water, questions emerge over who works together, exactly what we organise together and how we do it; in other words, it raises questions of governance.

In the context of this book, 'governance' is taken to mean '[t]he combined processes of defining, assessing, developing and implementing strategies to address collective problems which take place in overlapping arenas including research and science, public discourse, companies and government' (Voß *et al.*, 2006: 9). Governance, then, incudes processes of top-down government, but also the more complex and sometimes 'bottom-up' and 'networked' processes of public discourse, company decisions and actions, and research – all of which are done with (at least) some claim of serving collective needs. People and organisations contributing to governance include not only those in official 'government' positions, but also companies, voluntary groups, members of the public and researchers because all of our combined actions influence what is done. Governance includes the ways that people's thinking is shaped (concepts) as well as the formal and informal processes of organising collective actions (Figure 2.5). Studying governance, then, involves unpicking a range of these different influences on what is done in relation to a (potential) collective issue. Exploring the governance of water supply means examining who is supplying water and who is using water, where, when, how and why.

The historian Karl Wittfogel first coined the term the 'hydraulic state', referring to a paradigm in which control of water resources became a symbolic of a modern state (Wittfogel, 1957). While the formation of a state hydraulic paradigm in the UK through the nineteenth and early twentieth centuries did not threaten the country's 'very existence', as Scott Moore asserts it has done in Yemen, Moore's argument that in Yemen it 'destroyed traditional systems for water management and failed to construct a durable, equitable structure in their place' can perhaps be justifiably applied to the UK experience (2011: 39). The twentieth century is widely perceived as a golden age of the technocratic state hydraulic paradigm (Bakker, 2004). But while centralising and increasingly technical water services were highly esteemed in the nineteenth and early twentieth centuries, today's UK water management is heavily criticised: as too centralised; for dependence on costly technical expertise and hard infrastructure; as disconnected from the needs of particular micro-scale environments; and as both environmentally and financially expensive. In seeking to explore what forms of distributed and user-organised services might replace these technocratic systems, surprisingly little attention is given to highly relevant past experiences of water services which *were* locally distributed, which *did not* rely on costly engineering expertise, which *were* closely aligned with local environmental needs and which *were not* environmentally or financially expensive.

The narrative strategy of this chapter is to use two case studies of the same place but in very different periods. I begin with the story of water supply development in London in approximately 1500–1750. In particular, I examine the shift from dependence on in-situ water supplies obtained largely from common public access points to the widespread use of private piped water into homes. I then continue with a discussion of the contemporary water supply dilemmas faced by London. My historical sources of information include popular histories, as well as academic books and articles, in particular drawing on a new wave of environmental histories that explore the role of water and other socio-technical goods as societies developed, emphasising specific closely entangled connections that historically linked the 'social' with the 'natural' resources and systems. For the contemporary case study, I draw on Thames Water's documentation, as well as information about water supply options from academic sources around the world. Towards the end of the chapter, I use the discussion of water supply to elucidate some important dimensions of governance, which are equally relevant to the other collective issues discussed in this book, namely the management of water qualities, drainage and flood control.

From conduit to piped supply

It has long been recognised that water is a crucial factor in determining the placement of human settlements (Hoskins, 2005). Precisely how water was accessed, however, depended on the local hydrology and geography, as well as available skills and technologies. Drawing on long-established and locally specific environmental knowledge passed down through each new generation, options could include surface water sources such as springs, rivers and lakes, wells or pumps for accessing groundwater, rainwater cisterns collecting from roofs or yards, or conduits carrying water from springs into 'fountains' or water troughs.

In the 1500s, London was the most developed city in the UK, with 120,000 residents (Tomory, undated). Around this time, in different parts of the capital, all potential local sources of water were drawn upon; for example, Jenner writes of a housewife and a brewery both seeking to draw water from wharves on the River Thames (2000: 250 and 251), of the Thames being used for laundry, of many parish wells (ibid.: 252) and also wells in the homes of wealthy gentlemen (ibid.: 251), of cisterns used to collect and hold the rain (ibid.: 259) and of many conduits and standards for people to collect their own water, such as that provided from springs near Paddington to the conduit in Cheapside (ibid.: 253). The different sources came with different rules and expectations about use and abuse at a time when neighbours' fulfilment of their mutual obligations were central to social relations and the overarching concept of the 'commonweal', or common wealth to which all contributed and from which all benefited (Skelton, 2016). Rainwater collected from roofs was effectively 'owned' by the household, and likewise water from private wells. The provision and maintenance of public wells, pumps and conduits relied on philanthropic beneficiaries and/or public subscription: in these cases, parish vestries, city mayors and aldermen, or the local Manor. For example, London's Mayor and aldermen had arranged for water to be piped from the Abbot of Westminster's land in 1430 (ibid.: 252). In London there were 12 conduits by 1630, each of which had a salaried keeper enforcing regulations about water access (ibid.). Large users such as breweries, as well as wealthy aristocrats, paid for 'quills' from conduits to their premises, and their fees helped to support the system's maintenance (Van Lieshout, 2013).

People queued when collecting water from a conduit. Conduits were, therefore, known as sources of gossip and potential strife; for example, Jenner writes that, in 1561, a London water riot was only just averted when excessive use by aristocrats caused the Fleet Street conduit to dry up. However, collection of water from conduits was also associated with woman, younger apprentices, household servants and children – the least powerful members of households, who were most frequently tasked with water collection (Jenner, 2000: 254). There was a guild (or association) of water bearers who collected water from conduits and delivered it those who would pay them. However, Van Lieshout (2013) and Jenner (2000) both stress that arrangements for water collection and delivery involved moral and social as well as commercial connections; this is illustrated by the case of John Banck, who wrote in 1616 about maintaining a poor man through paying for him to collect water from the conduit (ibid.: 259).

'Urbanisation', or the growth of urban settlements, involves an increase in population densities as well as a spread of the urban envelope. In the case of London, the 120,000 population of 1500 rose to 200,000 by 1600 and to 500,000 by 1700, accompanied by the expansion of the urban envelope from the medieval city to create new suburbs, with the once walled medieval city of London merging with the previously separate upstream community of Westminster (Tomory, undated). These developments put an enormous strain on local water supplies both because there were more people whose needs were being serviced from finite supplies, and because of the potential for and likelihood of these sources becoming contaminated with urban sewage and waste. Early modern (1500–1800) responses to these challenges involved a mixture of protecting and rationing current supplies, as well as seeking to expand the supply envelope.

There were long-established rules about the protection of the quality of water sources in most communities. Dirty water was largely disposed of in sewers leading to rivers or directly into rivers, and rules prevented other behaviours that might threaten the purity of water supplies such as the deposition of solid waste such as manure or butchery waste. For example, residents in Sheffield were forbidden from washing clothes or calves' meat within 3 yards of a well; likewise, regulations limited the disposal of water from laundry into surface water streams, rivers and lochs that also acted as drinking water sources in Lanark, Edinburgh and Dunfermerline (Skelton, 2016: 22). Evidence of attempts to protect water supplies in London include the conversion of many London wells into pumps during the sixteenth century; these provided more protection from contamination (as well as removing the risk of falling in) (Jenner, 2000: 252).

There were also rules about the quantity of water used as well as the quality. In London a law from 1343 banned brewers and fishmongers from drawing water from the Cheapside conduit because they denuded the public supply (ibid.). Aristocratic households could have their quills (or private pipes) from conduits cut off if their households were understood to have been using public water excessively, as the Earl of Suffolk found to his cost in 1608 (ibid.: 253–4). Likewise, concerns about water shortage are likely to have stimulated the vestry of St Martin's, Ludgate, to limit the water collectors permitted to access the parish pump to just three in 1583 (and to just the men, not their wives or children), requiring also that they should deliver only to residents within the parish (ibid.: 252).

The expansion of supplies in early modern London included a law in 1543 that the city should have access to all springs within a 5-mile radius, as well as numerous examples of new and deeper wells. Supplies were extended in the 1570s–1610s through a waterwheel-driven pump under one arch of London Bridge (Halliday, 1999: 20), a horse-drawn engine extracted water from land by the Thames at Broken Wharf (Jenner, 2000: 257) and a canal (the 'New River') from Hertfordshire into the city (Van Lieshout, 2013). In all cases, entrepreneurs supplied some of the water they obtained via public conduits for general use, but their finance relied on the provision to private properties by pipe on subscription (ibid.) and, in this respect, they are the precedents a the long process of change described below. As Jenner (2000), Tomory (undated) and Van Lieshout (2013) point out, during the course of the seventeenth and eighteenth centuries, London underwent an unprecedented transformation from the city largely reliant on public water supplies or private wells/cisterns to one in which as much 30–40 per cent of the households within districts with piped water were enjoying piped private water supplies. The step change from the previous situation was two-fold. First, the investment required was huge. Though aqueducts carrying water to monasteries and public conduits had long existed, here the expense of collecting water from far away was increased by the costs of inventing, constructing, maintaining and administering technology and institutions of a distribution network. Second, what we would now call the 'business model' was different. Whereas previous investments had provided collective water through philanthropy, spiritual devotion or the institutional power of the church, the new system relied on commercial investors expecting a return on their capital through household payments for water.

The London Bridge water company initiated the use of private finance to deliver a combined public and private service; the latter focused on the water-hungry fish-monger trade (Tynan, 2002). The model was copied on a similar scale by the Broken Wharf company, but then on a much larger scale by the New River company. The latter tapped springs (and later the River Lea) at a significant distance from the capital in Hertfordshire, carrying the water 40 miles via a 10-foot-wide contour-hugging canal to a reservoir on Islington Hill. From the reservoir, elm pipes brought water into the city and used gravity to distribute it along streets (Tomory, undated). This gravity-driven system was able to have a much wider reach than those relying on the power of pumping. Originally financed by private city shareholders, led by the goldsmith and engineer Hugh Middleton, when financial challenges and obstructive landowners threatened the project in 1610 the King stepped in, doubling the available capital and allowing the project to be completed by 1613 (Stone, 2011; Jenner, 2000).

Household purchase of the 'right to access water' was an unusual transaction for the period as, unlike transactions carried out in a marketplace, there was no real-time exchange of goods for cash. Parallels were made between the right to access water with renting a property, so the New River company discusses its 'tenants', whose 'lease' required an initial bond and then a quarterly fee (Tynan, 2002). Although the costs per unit of water available were less than water purchased from water carriers, the upfront payments were such that only wealthier Londoners could afford it (Jenner, 2000: 258). The process of household provision involved valves operated by 'turncocks' controlling supplies to street service pipes that would typically flow continuously for a couple of hours two or three times a week. During this period, the water would fill a household's basement reservoir for later access by servants when needed, with any surplus water sometimes stopped by manual valve in the house, but sometimes flowing away (Tomory, undated). The use of a household reservoir or cistern mimicked the previous practice in which collected water was stored in the household until it was needed.

In its early years, the New River company struggled to attract sufficient 'tenants'. Just 350 took their water in 1613, rising to 1,035 in 1617 and then hovering around the 1,500 mark until the mid-1630s (Tomory, undated), the same time as the company returned its first profit (Hansen, undated). Probably because of the King's and city's investment, pressure was exerted on suitable households to take up the water; for example, Jenner (2000: 259) quotes a fascinating letter from (the previously mentioned) Mr Bancks to the local aldermen explaining his refusal to become a 'tenant' based on the age of his house, his capacious rain tank, and his employment of a local poor man to collect conduit water. Jenner illustrates that, as late as 1661, there was still a strong discourse resisting the idea of paying for water, citing a speech by a William Bell in front of the Mayor in which Bell approvingly links good judgement to the free flow of *conduit* water, 'not to your new River Water, that is imparted to none but those that will pay for it' (cited in ibid.).

The significant change in the uptake of piped water came after the 1660s. The New River company's customers rose from 10,000 in 1670 to 17,000 in 1683, and by the eighteenth century they were serving between 30 per cent and 40 per cent of houses in the expanding areas where they provided supply (Tomory, undated). A rash of six new water companies joined the three original suppliers during the period 1670–95, and a further six over the course of the eighteenth century (Van Lieshout, 2013).

One important factor driving this change was the Great Fire of London (Jenner, 2000; Tomory, undated). The fire destroyed the city's conduits and the piped infrastructure alike, but while both were replaced, the conduits were on a smaller scale than before. Meanwhile, demand for piped water supply increased because reconstructed buildings were able to include a basement cistern for piped water in a way that pre-existing properties had not. A second important factor driving the expansion of piped water supplies was the growth of the city. Tomory (undated) stresses the fashion for the gentry to purchase new well-equipped homes in the West End. Van Lieshout (2013) emphasises that, for the new Chelsea water company, particular marketing opportunities arose in new developments that were geographically distant from the Thames, and hence where residents would have more distance and trouble to obtain water. A third factor is that social norms among upper- and middle-class Londoners were changing. Whereas in the early 1600s water supply was a top-end good, only accessible to the very richest, by the late eighteenth century the presence of increasing numbers of middle class consumption-oriented town-dwellers meant there were many areas where the majority of households were on piped supply (Tomory, undated). According to Van Lieshout (2013) there were parts of London where it became hard for landlords to rent out properties that did not have a water supply.

The expanding networks created new challenges for the water companies in managing supply and demand. The long-term strategy to address this was always the access to more and larger supplies – beginning the 'predict and provide' tradition that we can see maintained up until the end of the twentieth century. However, technical innovations also enabled the companies to do more with their existing water supplies, hence offering temporary solutions to supply–demand imbalances. Key strategies included: developing expectations that no homes were connected onto mains pipes, just to street-level service pipes with less frequent flows; more careful surveying of elevations and management of new elevation-based water districts; new variable valve sizes to manage pressure; and the insistence that all homes had a stopcock to control the flow when cisterns were full (Tomory, undated).

In the late 1700s, some of the new water companies began using iron pipes rather than hollowed-out elm trunks that had required replacement every 30 years or so. This innovation significantly reduced leakage as well as providing the potential for water to be supplied at a higher pressure, and hence for longer periods.

It is important to be aware, however, that for a long period piped water supplemented but did not supplant other sources. As Van Lieshout's work shows (2013), direct access to the river was one ongoing viable source for household water, usually supplemented by rainwater tanks and the continuing use of public pumps. Johnson (2008) writes of Soho in the mid-nineteenth century demonstrating that most of the artisanal residents were collecting their water from pumps in this period (see the discussion of water quality in Chapter 6). So, while household water supply became a middle class aspiration, many other households – including the poor majority – continued to draw on public sources of water.

A number of general points can be made about London's water supply transition in the period 1500–1800:

- The advent of piped supply enabled demand to grow in two different ways. First, as Chapter 4 will explore in more detail, for any one household, the relative ease of accessing water associated with piped supply is likely to have led to wider and more profligate water use, and hence to have been associated with an increase in what we now call 'per capita consumption'. Second, the provision of piped water supplies enabled housing to be developed in locations that were not well served by surface water sources.
- What occurred was a partial transition. While piped supplies became significant and important, they supplemented rather than supplanted other sources of water. We need to wait until the early twentieth century before the 'transition' comes close to completion. Because the growth of piped supply coincided with (and contributed to) the process of urbanisation, the quantity of water accessed from traditional sources may have remained similar, or even increased, during this period.
- New processes of water distribution not only required new technologies, but also the new institutional practices of 'renting' a water supply. Significant improvements in the distribution technology also occurred through time with better surveying, closer management of quantities and (eventually) an improved quality of pipes.
- Driven by this expanding market and in a context in which the technical process of delivery (rather than the innate quantity of water) was seen as the key constraint, the advent of London's piped supply also began a process of 'predict and provide', in which companies sought to bring on new water supplies to meet anticipated increases in the quantity of water required.
- Although, on a city scale, the transition was partial, for specific households the water supply situation is likely to have been more binary: either they drew on the mixture of traditional supplies, or they had the new piped supply. In this respect, ironically, the process of connection to a water supply can be seen as a

first step in disconnecting people from more tangible engagement with water. Specifically, the potential for collective understanding of water supply problems and how to address them that arose around the conduit or the pump was displaced by a series of relations with a water company. Initially, there was still a strong collective element to this as people banded together on a street to request a supply or to complain about quality or quantity, but as customer service standards evolved such local collectivism diminished. This process of disconnection also involved the shift from an understanding about many waters of different qualities suitable for different purposes, to an expectation of one universal and high-quality good, water.

- Official community leaders like the monarch, mayors or aldermen permitted the change to piped supply to occur, by licensing or supporting new operators, but they did not perceive a public responsibility to initiate the system or to maintain the infrastructure. Some individuals (notably King James VI and I and Hugh Middleton) who had official positions did drive these changes, but they did so in their role as investors and business people not as community leaders.

To what extent are London's experiences as the pioneer in piped water provision paralleled in other cities and countries? Many of the described aspects of Londoners' experiences accessing water before the establishment of widespread piped supplies are likely to have been paralleled elsewhere and in other eras (and, indeed, may be paralleled in cities of the emerging economies today). For example, Geels' (2005) account of nineteenth-century Amsterdam and Patel's (2008) account of Toronto in the same era both stress the range of water sources used, their variable qualities and quantities and the how urbanisation led to increasing concerns about their quality and sufficiency.

In terms of the process of expanding supply, like in London, most local authorities welcomed the developments but they did not drive them. For example, early eighteenth-century Sheffield, an industrial city in the North of England, saw the construction of consecutively more elevated and distant reservoirs from 1712 onwards. Though the first venture was supported by the town trustees (a committee of local business people under the oversight of the Manor of Sheffield), subsequent reservoirs were built by entrepreneurs working for profit (Hey, 2010). Geels' account of water supply transitions in Dutch cities highlights the lateness of action on this front compared to London – mid-nineteenth century rather than in the 1600s or 1700s. The barrier, he suggests, was partly the size of the financial outlay needed to lay pipes from new water sources, but also because investors doubted the extent of returns: 'many people could not imagine that there was a profit to be made from such an ordinary product as water' (Geels, 2005: 381). He explains how most local authorities wanted investment to occur but not to make it themselves.

There are examples where local authorities did take the lead in developing water supplies, but not with aim of improving public health. In Toronto, for example, the city council invested in the services of a private supplier but did so for the purposes of serving fire hydrants, not for public supply. This meant a supply of low quality that few residents wanted to purchase (Patel, 2008). Similarly, the one exceptional city in Geels' account of Dutch cities is Maastricht, where the city council led the provision of piped water, but only as a side-product of a new system they had developed for flushing the canals (Geels, 2005). In a move that might be seen as contrary to the interests of maintaining equity and public health, a significant deterioration in the cities' maintenance of public pumps then helped to drive the development of the local authorities' commercial piped supply market.

There are also examples where city councils actually opposed the expansion of supplies that might help their populations. In the mid-nineteenth century, the municipal authorities of both Wakefield (a northern industrial UK city) and Cheltenham (a southern Spa town) blocked private investors from expanding the city water supplies because the sources that the entrepreneurs were intending to access were seen as threatening the well being of local industry (Hamlin, 1988). In Wakefield this action led to the continuing use of a contaminated water supply, which was probably a significant driver for the cities' exceptionally high mortality rate.

In this 'fly-by' account of water supply stories from other cities, it is clear that London has a role as the pioneer, largely resulting from the serious challenges of supplying water to its expanding population; out of pragmatic necessity it forged the model for both the technical and institutional processes of piped water supply. Nevertheless, and despite the different eras in which piped supply was introduced, a common factor between London and these other cities concerns the role of city authorities. It is clear that, until the end of the nineteenth century, the provision of water was not seen as a primary duty or concern of governance organisations. Indeed, insofar as community leaders influenced water management, they enabled or blocked water supply, with a focus on its impact on their cities' economic well being and relatively less concern for the health and well being of their residents.

Sourcing London's water for the twenty-first century

We now jump some 250 years forward into the contemporary era. Just as the water supply for a growing London was a concern in 1600, so it is today. Just as the options for finding more water from the current (in-situ) sources were being exhausted in 1600, so the current prospect of continually increasing abstraction from near and far water resources is limited. Just as entrepreneurs innovated, looking for new solutions in the late 1500s, pioneering the transformation of the water system, so there may be a need for yet another new and different, but similarly disruptive, transformation today. As in the historical example above, the aim in the contemporary case study below is to use the exemplar of London to illustrate some of the opportunities and the challenges in relation to the governance of water supply and demand. Brown and Farrelly (2009) and Wong and Brown (2008) highlight the need for utilities to build resilience to drought by drawing on diverse sources of water. Each source has different reliability, environmental risk and cost profiles. If a variety of sources is available, they argue, utilities are able to manage the sources they draw on in real time to achieve the best balance of environmental cost, risk and expense. Wong and

Brown (ibid.) and the International Water Association (Binney *et al.*, 2010) additionally strongly emphasise the need for 'fit-for-purpose' water: the norm that has dominated the twentieth century in which water is treated to potable quality and distributed for a wide variety of future uses is called into question. Could we be more selective, they ask, saving energy and water by treating water only to the standard required for its use? In appraising London's developing options for its future water supply, it is useful to consider both the variety of future sources and the extent to which they move towards 'fit-for-purpose' supply.

London's contemporary water system is the legacy of that which has developed in the past. The piped water supply is the universal source of household water *and* the source for most commercial users of water. The pipes and supply infrastructure are run by a private monopoly, Thames Water, covering the whole of London and other parts of South and East England, serving 9 million people, which is anticipated to increase to 10.4 million by 2040 (Thames Water, 2014a: 4). Customers are spread through four different water supply areas (see Figure 3.1). Of these, the greatest pressure is on the London supply area, and this is the focus of some of the detailed discussion below.

Thames Water currently supplies about 2,500 million litres of clean water per day (Ml/d). Pressures on Thames Water's water supply system result from a combination of what they see as 'demand' and 'supply' factors. Demand, incorporating household use, commercial use and leakage, is expected to increase by about 10 per cent, or 250Ml/d. This is largely the result of the growing population. Water supply is achieved

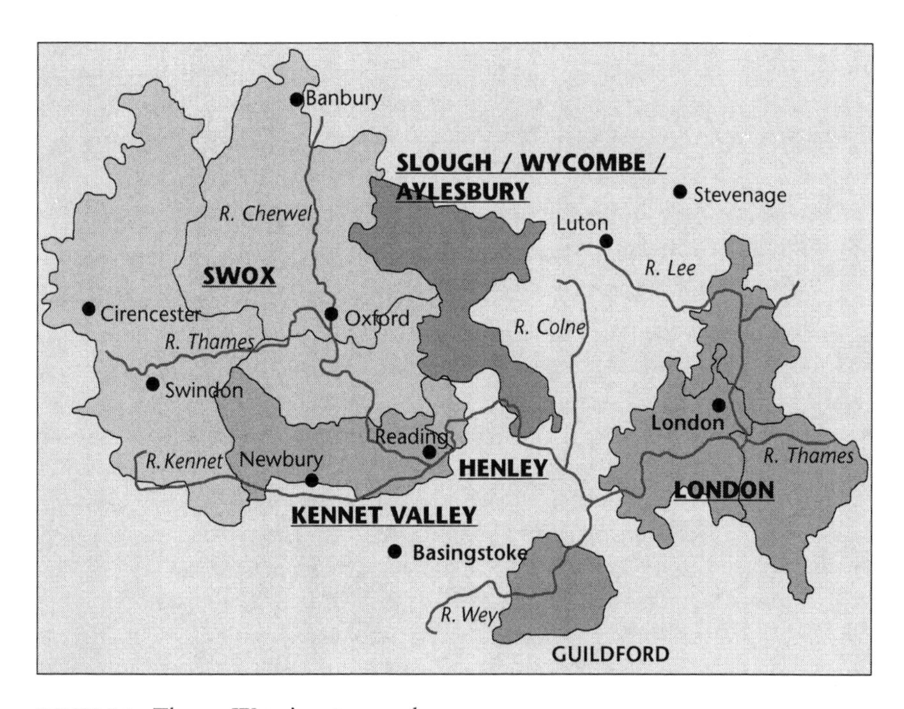

FIGURE 3.1 Thames Water's water supply areas
Source: Thames Water, 2014b

largely from surface water abstraction supported by storage reservoirs and some groundwater abstraction. The implementation of the EU's Water Framework and Habitats Directives means that the water company needs to reduce its surface and groundwater abstraction by about 21Ml/d by 2040. Meanwhile, predicted changes in climate mean that there will be less water available, further reducing the amount that can be abstracted by an estimated 90Ml/d (Thames Water, 2014b). The anticipated increase in demand, combined with the decrease in supply, will lead to a 360Ml/d difference between the water available and that needed in 2040. Thames Water's *Water Resource Management Plan* describes how the company intends to address this, presenting what they view as the 'combination of schemes that solve the supply–demand deficit at least long term cost' (Thames Water, 2014b: 20).

Of course, these figures are not value free. As the baseline numbers in Thames Water's final 2014 *Water Resource Management Plan*, they are the outcome of complex processes of modelling of both supply and demand, with model assumptions and processes then contested by regulators and the public through the development of the draft and final plan. Inherent assumptions about, variously, climate change, hydrology, the performance of assets and (particularly) the behaviour of the public are the subject of disputes between parties with an interest in this highly political document (Walker, 2013). The results of these contests matter because they influence how the environment will be treated, what investment funds are available and where they will be directed, and who/what will be seen as responsible if water shortages do occur. They can also have a serious impact on public health; for example, if households begin to receive a separate water supply for purposes such as washing a car, this substandard water could potentially be ingested by a minority of water users who could become seriously ill.

Notwithstanding these complexities, the discussion below reviews a variety of responses to this challenge. The responses discussed can be understood in three sets. The discussion begins by appraising traditional responses to water scarcity: drawing water from yet further away, increasing local connectivity and increasing storage. It then considers three mainstream modern approaches: the development of desalination plants, addressing leakage and managing household and business demand. The last set of responses includes four, more novel approaches that are used elsewhere but not yet in use in the London area. These are: the reallocation of water between uses; water reuse; and de-centralised household/community rainwater/grey water use. The section concludes by considering processes governing decisions over water supply and demand, and whether changes in the governance structure have the potential to help address future water management challenges.

So, beginning with traditional responses, droughts in the southern and eastern part of England frequently raise questions of a water grid, and whether bulk transfers from wetter parts of the UK might be a viable way to address future water resource questions. Such transfers would certainly be technically possible and are in line with the long tradition of areas of water demand collecting water from increasingly distant locales. They would be large, technical and highly engineered interventions using increased supply to address supply–demand balance issues, and hence can be seen as the epitome of a 'technocratic' solution. Between 1975 and 1982, for example, the Northumbrian Water Authority (now Northumbrian Water Limited) connected North-East England's River Tyne, Wear and Tees to the UK's largest manmade lake, Kielder Reservoir, via a subterranean aqueduct which enables the 'Kielder Scheme' to be managed essentially as a regional water grid. In the London context, there are two significant objections to bulk transfers between entire regions. First, there is an environmental concern. The regions that are judged to have spare water capacity are the most distant: in the far North and East of England, for example, or to the West of Wales (EA, 2008a). Transferring untreated water from distant river basins has the potential for mixing waters of different aquatic qualities, with potentially negative ecological impacts. Removing water from a formerly 'wet' region also has an environmental cost and is liable to change the local ecology. Second, water is a heavy object; providing and running the infrastructure to allow bulk transfers is not a cheap proposition. In particular, pumping water over hills is very expensive. For these reasons, in 2006 the House of Lords Science and Technology committee concluded: 'a national water grid is not feasible' (House of Lords, 2006: 167). The UK's decision on this issue is in line with current international thinking. The World Commission on Dams has objected strongly to the use of bulk water transfers as a first solution to water supply problems, without sufficient consideration of alternative solutions. It has also highlighted the poor governance associated with such projects, giving insufficient attention to concerns raised in both 'giving' and 'receiving' communities (Imhof *et al.*, 2002; International Rivers, 2016). The World Wildlife Fund likewise stresses how bulk water transfers are financially expensive, how they alter hydrology and contribute to desalination (WWF, 2009).

Another solution with a long legacy is to increase the connectivity between adjacent water resource regions. Drawing on Californian experiences, Nevarez comments that interconnections allow demand to expand beyond local natural limits (Nevarez, 1996). Taylor and colleagues highlight how the drought of the 1890s prompted London authorities to bring the different water supply companies together, leading to a new united and municipal water supply organisation from 1902 (Taylor *et al.*, 2009). They comment that the impact of these interconnections was to reduce local vulnerability and to allow further expansion of demand, but that when later drought hit the new interconnected context, it was felt across a broader area. In the contemporary era, various UK governments have asserted a similar pressure to their London predecessors, calling for better coordination of water resource management across the water companies in the whole of South East of England. A 'Water Resources in the South East' group now operates, seeking to identify water resource solutions, including those that cross water company boundaries, and hence to inform companies' own processes of resource planning. In a review of possibilities for the next 25 years, the group identified over 1,000 options for addressing water resource deficits, about a quarter of which involved water transfers between water companies in or adjacent to the South East (WRSE, 2013). Thames Water's *Resource Management Plan* demonstrates their participation in this; specifically, the most water-stretched resource zone in Thames (London) is signed up for exporting water to the even more resource-strapped Affinity Water, with quantities increasing over the planning period (Thames Water, 2014b: 15). Transfers of water into the Thames region from the adjacent Severn basin also feature as part of the long-term planning (Thames Water, 2014b). In relation to the London water resources area, the preferred programme of water resources to 2040 includes 11 per cent of new resources obtained from water transfer (ibid.: 15).

The third 'traditional' solution involves increasing storage in order that supplies can be maintained in times of water scarcity. This is an intervention that has been given serious consideration by London in the form of proposals for an 'Upper Thames Reservoir' covering 4 square miles on farmland south of Abingdon in Oxfordshire. Initially suggested in 1990 (Garis *et al.*, 2003), the proposals formed part of Thames Water's draft *Water Resource Management Plan* in 2009. The proposal was rejected in 2011 by the government's inspector, who ruled that the plans did not fit the available evidence and that viable alternative water resources were not fully investigated. Nevertheless, the reservoir remains as one potential strategy under consideration for addressing long-term differences between water supply and demand for London (Thames Water, 2014b), though it is not in the list of preferred options identified in 2014 (Thames Water, 2014a). Nevertheless, the local council continues to 'safeguard' the land from other developments in case reservoir development is approved (Vale of the White Horse, 2016). Meanwhile, the website of the Group Against Reservoir Development remains active (GARD, 2016), and the debate rumbles on, with a January 2016 report from the Institute of Civil Engineers arguing that the reservoir is 'inevitable' (*Oxford Mail*, 2016).

Moving from traditional solutions to more contemporary technologies, one highly technological means of creating potable water resources involves the desalination of sea or brackish water to achieve potable standards. Desalination is an attractive part of a diversified portfolio because it is less dependent on the weather than most other sources. However, the capital and running costs of desalination plants are high, and they have environmental implications in terms of their high energy demand, combined with the challenge of how the residual concentrated brine is disposed of (Water in the West, 2013). An example of the use of the desalination elsewhere is discussed by Richter and colleagues (2013), who highlight how Adelaide's new desalination plant is going to supply 25 per cent of the city's future water needs, significantly reducing the city's reliance on water from the Murray Darling basin. However, the plant is largely responsible for a 400 per cent increase in water prices. London's first desalination plant opened for potential business in 2012 in Beckton, on the Thames estuary, aiming to produce water of potable quality from water in that estuary. By focusing on the treatment of brackish water rather than sea water, the Beckton plant minimises energy usage. The Beckton treatment works provides a back-up water provision in case other sources of water are not available. However, like all desalination plants, the energy demand is high – five times that of other forms of water treatment (Bell, 2015). Although Beckton is run for a period every year to check that it is functional, it has not yet been used as a back-up source of water because London has not been subject to a drought since the works opened. The Beckton treatment works was the subject of political controversy, with planning consent initially rejected on the basis of its energy demand, only for this decision to be changed when the London Mayor changed from labour to conservative (ibid.). While the Beckton plant will remain operational for some decades, plans for London's future water resources do not envisage any further development of desalination.

A quarter of London's cleaned water is lost through leakage (CCW, 2015b). Such leakage is not surprising. London's water supply travels through a myriad of pipes, many of which date back more than 100 years; it is to be expected that such pipes will have many fissures and small holes through which pressured water can be lost. Although the UK convention is to frame the leakage of cleaned potable water from supply pipes as part of 'demand', in practice this is a misnomer. The name comes from the engineering perspective that the leakage is part of the water 'demanded' from the reservoir. However, in everyday usage, the 'water demand' suggests that there is someone with a real and present need for water, which is certainly not the case with leakage! Leakage quantities are directly important to the supply–demand balance because they are a significant loss from the potential supply available. But leakage also plays a second, indirect but symbolically important role. The water supply situation comes particularly into public scrutiny during periods of drought. Customer research underlines that leakage is seen as important: a perception that leakage is being addressed can make it much easier for utilities to encourage those they serve to save water (Chappells and Medd, 2008; *Utility Week*, 2016).

To what extent is it appropriate for water utilities to take action to prevent leakage? The UK's economic regulator, OFWAT, has developed a measure called the 'sustainable economic level of leakage' (SLL) (OFWAT, 2016) that evaluates the economic and environmental costs of a leak against the financial costs of fixing it. The measure recognises that not all leaks are worth fixing; but does the sustainable level of leakage address enough? A key point is that this is concerned only with the direct benefits of fixing a leak, and does not focus on the indirect impact on public communication and understanding. For a while, under the Labour administration, OFWAT applied 'leakage targets' that were stricter than the SSL, which drove innovation in the industry for leakage to reach an all-time low in 2011 (*Utility Week*, 2016) – also delivering innovation benefits for the UK industry, which has led the world in leakage detection (Speight, 2015). Despite clear evidence that customers remain concerned about leaks, leakage rates have crept up since 2011 (*Utility Week*, 2016). During the asset management plan (AMP) period 2015–20, Thames Water aims to further reduce leakage by 59 million litres per day, to be stable about 10 per cent less than current levels – somewhat more effort than required by OFWAT. Much of this is to be delivered through the benefits of requiring more customers to be metered (see below), which will aid the process of leak detection. A question remains about whether such action will be seen as sufficient by the people of London if a drought restricts future water availability.

Seeking to manage household demand for water forms a significant component of Thames Water's plans for addressing its water shortfall during the period 2015–30, marking an important shift from the previous 'predict and provide' approach to water management (Thames Water, 2014a). The main means of achieving this demand reduction is through the combination of progressive metering and water efficiency. Progressive metering means introducing metered charging to all properties within a district. The 'progressive' (i.e. compulsory) nature of the programme is a significant change to the previous default approach in the UK in which companies made meters available to those households who opted for them, with the result being a rather ad hoc (and expensive) process of meter distribution and meter-reading requirements. The new programme began with a pilot in Bexley, installing 4,400 smart meters in 2014 (Thames Water, 2016a and b). The 'smart' nature of the meters means that they can be read remotely, enabling real-time monitoring of flows and changes in flows, hence supporting the process of leak detection. Households with meters installed under this new scheme have two years' 'grace' to adjust to metered charging, though they can choose to switch earlier if they want. Accompanying the progressive fitting of meters, households are offered a visit from a water efficiency advisor who can also provide and fit water-efficient devices. In recognition that some households may struggle to pay their bills, households are eligible for a social tariff (with the potential to halve their water costs) if their water bill takes more than 3 per cent of their household income and if they are on welfare benefits or are in receipt of a low income with a vulnerable (that is, very old, very young and/or disabled) person present. The progressive metering scheme will roll out across London over the subsequent 15 years, cementing (at least in the Thames area) a long-anticipated change from fixed to variable water charges.

Moving on to the more novel approaches that are not yet much in evidence in the UK, water 'reallocation' involves the reassignment of water rights from traditional sectors (such as agriculture) to newer uses, where it is valued more highly. Such reallocation or water trading is important to include here because it is a significant source of water for some drought-stricken cities. For example, Richter and colleagues (2013) highlight the potential for conservation in irrigated systems, alongside schemes to pay farmers to leave land fallow, citing examples in San Antonio and San Diego – where schemes have successfully secured more water for these cities. As the world urbanises, it is certainly appropriate to question the balance between urban and agricultural uses. However, in this respect the UK is in an unusual and almost unique position. While across the world 70 per cent of abstracted water is used for agriculture, in the UK it is as little as 1–2 per cent, as most agriculture is rain-fed. In this respect, the reallocation of ('available' or blue) water from agriculture to the London supply has nothing to offer. Indeed, if climate-induced drought limits overseas food production, it might rather be a concern that urban water use needs to reduce in order to support more irrigated agriculture in the UK.

In contrast, the idea of water reuse has gained currency and is considered to have substantial potential to support future urban water use. Water reuse involves treating waste water to such a high standard that it is suitable for other uses. In practice, a densely populated country like the UK has indirectly 'reused' water for centuries; it occurs whenever one utilities' water supply inlet pipe is on a river downstream of another's sewage outfall. However, while such indirect reuse can be unseen by the public, direct reuse in which sewage is treated for potable use without dilution in a river has long been assumed to be publically unacceptable. These assumptions have been challenged over the past few decades as water shortages have driven reuse schemes in a number of localities: for example, potable reuse has been in operation since 2013 in the Colorado River Municipal Water District (Texas Water Development Board, 2015). Pending further investigation, large-scale water reuse is Thames Water's preferred option as a medium-term solution to London's water issues (Thames Water, 2014b: 27). The emphasis on effluent reuse over desalination arises for obvious reasons: a proposed effluent reuse plant uses far less energy per unit of water (Bell, 2015). Initial plans indicate a reuse plant at Beckton will be operational from about 2027, when Thames Water intends: 'to monitor the first plant installed, and be confident in both its operation and acceptability to customers before committing to building another' (Thames Water, 2014b: 33). These plans for water reuse form 34 per cent of Thames Water's preferred options for delivery of new resources by 2040.

There is an interesting contrast between Thames Water's plan for Beckton and the reuse activities underway in Singapore, which has transferred the estuarine Marina Bay into a large reservoir for storm water recycling (Wong and Brown, 2008; Chew *et al.*, 2011), which is then treated for potable use. As will be explored in more detail in Chapter 6, the UK's combined sewage system mixes surface water with foul sewage for disposal. Such mixed sewage is more expensive and difficult to treat than either foul sewage or storm water by itself. However, the potential for large-scale reconstruction of London's sewerage system to separate foul from surface water is very low, as both the costs and the disruption would be enormous. In this context, London is extremely unlikely to emulate Singapore's system, despite its relative cheapness and higher level of community acceptability.

Although reuse is a relatively new technology that breaks new boundaries in terms of public acceptability, the proposed reuse of water at Beckton can, nevertheless, be seen as in line with traditional technocratic solutions: this is because it provides a single centralised technology, aiming to produce potable water for universal use. More radical variants of 'water reuse' generate water of a non-potable quality for a variety of local uses. Examples from elsewhere include:

- *Use of city as a catchment*: Spaces in a city can be used as a source of rainwater for non-potable uses. For example, in Melbourne, drainage in the city centre Olympic Park was reconfigured: rather than flowing into the River Yarra, storm water is collated, treated and stored for reuse in landscaping and toilets (Business Environment Network, 2011).
- *Effluent reuse for landscaping*: Community distaste for the idea of consuming used sewage, alongside the high cost of treatment to potable standards, means that a common policy in some arid environments is to use recycled water for specific non-potable uses such as landscaping or agriculture. Such non-potable recycled water makes up 4 per cent of the water supplied in San Diego, and 6 per cent of that supplied in Phoenix (Richter *et al.*, 2013).

- *Bulk reuse for non-potable uses via dual distribution system*: Alternative 'dual' distribution systems can distribute non-potable water for suitable non-potable uses. As Binney and colleagues (2010) stress, dual pipe systems, which deliver non-potable water alongside the standard potable supply, can be fed by a suite of local sources including recycled water, storm water or groundwater and might be used for a range of non-potable purposes such as toilet flushing, laundry, garden watering and open space irrigation. Dual pipe systems require processes of community learning: new norms are needed to distinguish potable from non-potable water (for example, non-potable pipes are coloured purple), and household plumbing takes a new and different form. The costs of retrofitting are such that dual pipe systems should be fitted into new properties when built, irrespective of whether alternative water systems are yet on-stream (Wong and Brown, 2008). Such dual distribution systems have been central in addressing water demand in the US city of San Antonio by 20 per cent (Richter *et al.*, 2013).
- *Building/community-level rainwater or grey water harvesting*: Reusing non-potable water in a building requires individual developers to fit rainwater or grey water recycling systems with a dual pipe system. The same systems can be implemented on a wider selective basis across different households/buildings with institutional support. For example, some Australian water companies have offered subsidies to support the implementation of household rain tanks, often with a mandated feed into household laundry or toilets (Sofoulis, 2015). Likewise, Woelf-Erskine discusses how Californian water governance has prioritised more decentralised water decision-making, which sometimes translates into support for rain tanks (Wolfe-Erskine, 2015). These subsidies have sought to relieve the pressures on water resources at seasonal pressure points, as well as offering the potential for reduced storm water flows. As Sofoulis stresses, many Austalian households have fitted rain tanks even without subsidies, demonstrating their concern for ensuring some independence from the potentially restricted municipal water supply (Sofoulis, 2015).

At present, London has very limited building-level examples of non-potable reuse. For an exceptional example, the World Wildlife Fund implemented grey water and rainwater recycling in their new UK headquarters building in Woking (Aquality, 2016). Likewise, with some support from Thames Water the Bangladeshi community's new mosque at Wapping includes a dual pipe system to enable water used for ritual washing to be reused for toilet flushing (Thames Water, 2016b). Wider processes of non-potable reuse have not been instituted, however. There may be a number of geographically specific reasons why London might be slower at instituting these processes of non-potable reuse than other water-stressed cities. For example, the legacy of the combined sewer system means that there is less pure storm water or foul sewage to be treated than in other cities with more consistently separated sewers. Moreover, London's temperate climate means that the demand for non-potable water for landscaping is smaller than elsewhere (though its demand for toilet flushing and laundry is likely to be comparable). Also, the density of London's development, combined with its relatively small plot sizes means that there is limited space for household-level rain tanks or reuse facilities, though a neighbourhood or community-level facility might be equally welcome. Beyond these geographically specific reasons, however, there are broader institutional and cultural issues that are likely to play a part in the lack of development of non-potable infrastructure. Reviewing attitudes to rain tanks in an Australian context, Sofoulis (2015) highlights the challenge they pose to institutional managers of water because they do not fit the normal centralised and technocratic pattern of water management. Although developed in the specific context of household-level rain tanks, many of Sofoulis' arguments seem to apply across the field of non-potable reuse systems.

One important element of non-potable reuse relates to the utility not being in control. As one of Sofoulis' interviewees commented (ibid.: 531) 'I can't imagine [the water corporation] being responsible for… all of these dotted-around-the-place little things', while other interviewees told 'dumb owner' stories, highlighting their worry that owners of rain tanks may not maintain them properly and hence may expose users to risks. (Similar risks are likely when householders have access to a community or institutional non-potable supply.) Sofoulis reflects that we may not be so much talking about 'dumb owners' as 'dumb design', stressing how, in practice, there is an institutional challenge concerning how consumer-maintained infrastructure (or, indeed, non-potable water infrastructure) can be designed and developed in a way that maintains owners' knowledge and skills relating to appropriate use, especially when properties change owners or occupiers. In a UK context, households enjoying a non-potable as well as a potable supply would certainly be a significant change, and processes of learning would need to take place. Nevertheless, it is worth noting that we have systems for managing the potentially lethal machinery associated with household electricity, or personal vehicles; is it not possible that similar systems can be instituted for the significantly less dangerous supply of non-potable water?

Another difficulty for water managers associated with decentralised technologies concerns their funding. From a utilities' water supply perspective, a rain tank or a non-potable reuse facility is, effectively, a very small reservoir. Were Thames Water to appraise subsidies or support for rain tanks or grey water recycling in their *Water Resource Management Plan*, they would almost certainly come to the same conclusion as the Australian 'Productivity Commission', which concluded rain tanks almost always represented a poor investment compared to other options for expanding water supply (Sofoulis, 2015). Such an assessment provides an appropriate appraisal of water supply costs, but ignores the other important functions that such facilities can perform. One important function in Australia (which would also be important in London) is their impact on surface water drainage (see Chapter 6); rain tanks or rain-using community reservoirs that are empty when rain starts effectively act as a buffer against flooding. Other significant factors relating to Australian and Californian rain tanks included supporting the local ecology through offering water for people's gardens during drought (Sofoulis, 2015), and their contribution to education about water management (Woelfe-Erskine, 2015). But as these considerations fall, respectively, outside the remit of the water resources team (surface water drainage) and the utility (greening), and/or outside the imagination of many water managers (education), they are excluded from water supply cost–benefit calculations.

Just as it is hard for the utility to justify expenditure on water tanks for themselves, it is also hard for water regulators and utilities to imagine that a consumer would sensibly invest in a water tank or other alternative system at either an individual or community level. Sofoulis suggests that these organisations assume that the consumer's decisions about such investments would be based on a financial calculation. As will be discussed in Chapter 5, this perspective on consumers is limiting and does not take account of other things they value. For example, evidence from London (Chappells and Medd, 2008), Australia (Sofoulis, 2015) and California (Woelfe-Erskine, 2015) suggests that people like rain tanks because they give them independence from water companies and the restrictions that they may make, as well as enabling contributions to local ecological well being. Without this broader understanding of householders' perspectives, water utilities seem destined to miss a trick in terms of household-level efforts for water recycling and reuse.

Finally, it is noteworthy that, from April 2017, a regulatory innovation may impact on London's water supply processes. Commercial customers will be able to choose their water supplier, just as domestic customers choose their energy supplier now. In many respects it is purely a superficial change in the market processes: a commercial customer who shifts from Thames Water to another supplier may get water from the same sources and through the same pipes as before. However, there may also be scope for more development and enhanced tailored support to customers about water efficiency. In Scotland, for example, where commercial customers have been able to choose their water supplier since 2008, a new water retailer has developed an arrangement with the National Farmers' Union to sell water to farmers across Scotland alongside providing a package of farm-oriented water efficiency support (Reynolds, 2016).

There are also plans to further develop this regulatory innovation, with a roll-out of competition to domestic customers, though it is not yet clear when this will take place. At the time of writing (summer 2016), questions are being raised about whether the small household savings delivered through such a roll-out would compensate for risks of competition such as decreased trust and potential risk to the resilience of supplies (Darcy, 2016).

In conclusion, it is clear that concerns over the environmental consequences of water abstraction and transfer have ushered in a new approach to the provision of urban water supply. The old assumption that all demand should be met from extracted resources has been swept away. However, the nature of the new water supply paradigm is not completely clear. In London, at least, it is apparent that a significant role will be played by the continuation of centralised one-size-fits-all provision; expensive desalination and water recycling plants offer predictable and controllable assets in which utilities provide water to a relatively passive population.

Household-level demand management also plays a role. Wong and Brown (2008) comment, however, that there is a risk that governments will overlook the importance of building a diversity of sources once centralised schemes, such as desalination or water reuse plants, are commissioned. Whereas some cities with water supply challenges have pushed beyond this point, operating dual supply systems at a city, neighbourhood or household level, such approaches do not appear to be under consideration in London. These approaches involve close partnerships between utilities and other organisations such as municipalities, as well as with the public, and hence perhaps present a step too far from the traditions of technocratic control.

Water supply governance

At the heart of this chapter is the governance question 'how should we obtain water for people in towns and cities?' As has been illustrated, a governance perspective raises three important sets of questions about how water is supplied.

The first question concerns which collective scale – in other words, with which 'we' – is the appropriate grouping concerned? Depending on status and household infrastructure, for sixteenth-century Londoners 'we' might operate at a household, neighbourhood, ward, parish or London-wide level. Hence, a household may have sufficient water in their rain tank, the neighbourhood might benefit from access to the Thames, or the city might require the provision of the New River's water. In contrast, for almost all contemporary Londoners 'we' is very clearly at a collective 'London' or 'Thames Water' scale, with very few commercial and virtually no household customers enjoying their own separate sources of water. Moreover, as we have seen, while a shift to a more hybrid set of 'fit-for-purpose' sources of water is underway in some cities, this is not really the case in the Thames area (or, indeed, in most parts of the UK). There is also a question of where 'we' (whoever that is) can draw water from and how far away it is. The physical scale from which we are technologically able to draw water has increased through time; but whether we exercise this ability depends on issues of governance. Sixteenth-century London may have been able to divert water from the River Lea, but for contemporary supplies to draw water from Scotland or Wales has been judged unacceptable. In the sixteenth century, royal consent was needed in the form of an act for large works such as the construction of the New River. In contemporary London, decisions about which sources are drawn upon involve multiple levels of governance, with significant input from both the company, Thames Water, and from the environmental and economic regulators (Environment Agency and OFWAT), all within the framework of European legislation.

A governance perspective also leads us to explore how water supply is combined with different functions. Water supply may be conducted alone (e.g. some UK companies) or with sewerage (the large UK companies), with drainage (e.g. in France, and in pre-1973 municipal systems in the UK), or, indeed, with a variety of other utility functions such as electricity and gas supply (e.g. in some French utilities), or canal flushing (the historic example of Maastricht). Those supplying water may also be charged with managing demand, or demand may be conceptualised as something external that simply has to be met. Water supply may be split into a wholesale component (collecting the water and making it available) and a retail component (distributing and selling it), as, for example, occurs in Melbourne, Australia; equally, the delivery of water may be executed by the same body that makes decisions about whether resources are to be exploited, or such regulatory functions may be carried out separately. Different understandings of collective priorities and needs prompt different combinations of functions. As we have seen, the contemporary separation of water supply from drainage in London makes it difficult to conceptualise the advantages of avoiding flooding and pollution by keeping rainwater at source, and saving on water supplies by providing it to households, in one calculation. Likewise, the separation of land use planning from water means that there is no attention to water shortage issues when considering the location of new homes. Clearly, there is a balance to be struck: complexity is created when decisions are considered from multiple angles, but reducing decisions about water supply to exclude ecology or flooding, for example, seems to belie the interconnections of the hydrological cycle.

The final, but probably most important and contentious issue concerning governance relates to ownership and management; most critically, this is concerned with *who* is making the various decisions about water supply, and hence (implicitly) whose interests they serve. This issue is discussed in more detail in Chapter 8, but here we can note that potential decision-makers include the state (or experts to whom they have delegated it), the market, or communities. Before the mid-nineteenth century, control over water resources varied, but was generally decided at the local level (Strang, 2004; Moore, 2011), with specific resources provided by the (effective) local state, and/or through commercial processes – for example, through water carriers. From the mid-nineteenth century, management of water commanded increasing attention from the state, and was allocated as a state resource in international law, in what has become called the 'state hydraulic' paradigm (Wittfogel, 1957; Bakker, 2004; Moore, 2011; Walker, 2013; Bakker, 2014). From the mid-twentieth century, this approach came under increasing question: it was seen as inefficient and unable to care for the environment or to invest in new water assets (Bakker, 2004; Walker, 2013). A new approach to water management, based on 'market environmentalism', sought to use the market to support investments in water assets while using state regulation to maintain the quality of water services and to improve the treatment of the environment (Bakker, 2004; Walker, 2013; Bakker, 2014). Currently, theorists such as Wong and Brown (2008) and Sofoulis (2015) are spearheading pressure for water utilities to connect the water cycle more fully through using waste water for water supply, doing so at more and multiple scales, and to work more imaginatively with the people living in the catchment they serve.

In practice, as Bakker (2014) stresses, debates over state and market control of decision-making function along a range of dimensions (see Table 3.1), which are present in different countries to different extents. In this respect, while market environmentalism has continued some controlling aspects of the state hydraulic paradigm, there are great variations between and within states in relation to asset ownership, institutional management, decision processes and local charging for water services. For example, many countries – including Australia, the USA and most continental European countries – have long operated processes of metered charging at the domestic level. However, such apparently marketised charging systems sit alongside municipal ownership of water services. Bakker's different dimensions of water governance provide a useful framework through which to review the changing governance of water services in London through the two periods considered (see Table 3.1). What is clear is that seventeenth-century London delivered a very mixed economy, with assets owned, managed, exchanged and valued in a variety of different ways. In contrast, contemporary water users in London are almost all in the same situation of dependence on a single, monopolistic water utility, attentive to the environment in

TABLE 3.1 Dimensions of privatisation (Bakker, 2014) and their presence in historical and contemporary London

Dimensions	*Seventeenth-century London*	*Contemporary London*
Ownership of resources, storage and distribution assets	Assets owned variously by households, parish authorities and private companies	Assets owned almost exclusively by a private company
Management of organisations collating and distributing water	Water companies and water carriers both worked commercially; but water delivery also part of the social economy for maintaining the poor, and access to conduits and river involve a mixture of collective decision-making and hierarchical diktat	Thames Water, managed through performance measures from regulator, with input from local consultation
How environmental value is measured	Environment is not explicitly considered	Environmental capacity provides absolute limits on water extraction for environmental regulator
How water is exchanged	Water is free at the conduit and in the river, but, at household level, water must be collected or paid for	Water (almost) only available from monopoly supplier at cost. Charges either fixed according to property size, or metered charging by volume used. Commercial customers choose water retailer from 2017
Input to critical decisions and action	Multiple stakeholders take decisions and actions at different levels in relation to different sources of water	Top-down decision-making by water supplies/regulator supplemented by consultation mechanisms. Householder perceived as rational money-maximising actor

respects required by the regulator (e.g. quality of waste water) or customer (e.g. the extent of leakage), but not innovating to deliver any diversity of water sources in terms of more distributed non-potable water sources that others have seen as marking out the sustainable city of the future (Wong and Brown, 2008; Binney *et al.*, 2010; Speight, 2015). Walker (2013) highlights some of the contradictions of this contemporary system, stressing the difficulties for an organisation that makes money out of selling water in also asking its customers to reduce their water use.

In reviewing the differences between water supply in pre-modern and modern London, we can see the changes wrought through technocratic water management through a particular supply-oriented lens. Variable 'waters' – from rivers, cisterns and other sources – have given way to a single universal (potable) water, drawn from increasingly distant locales. Whereas a variety of individual markets for providing and transporting water operated in the sixteenth century, in the twenty-first century there is a single, albeit highly regulated, supplier. Whereas sixteenth-century authorities dabbled in ensuring water availability through cisterns, rules about access to the river, or by encouraging investments, state regulation of water is a prerequisite in twenty-first-century London, occurring at a local, national and international level, and encompassing new imperatives concerning ecologically sustainable levels of abstraction and treatment. Moreover, whereas water supplies and management were a daily concern for the sixteenth-century resident, water commands little thought from most contemporary residents. While a sixteenth-century resident would be alive to quantities and qualities of water, individual innovation to collate water or to conserve supplies today is relatively rare. Interestingly, though, recent developments in Australia and other drought-ridden areas suggest some possibility of reinstating household water collection – not as a stand-alone supply infrastructure, but as a hybrid addition to centralised water provision.

Overall, the story of urban water supply is a significant part of the history of the technocratic model of water management. Urbanisation has been associated with a long process of modernising water – reducing the varieties of waters, centralising its collection and distribution, and distancing the consumer from its sources or management. Where evidence is available, it appears that these processes were more driven by economic and commercial benefit – and less by public welfare or health – than is generally understood. The idea of measuring and controlling water resources on a national level could also be seen as encompassing the technical optimism of the 'modern' era. Ironically, however, the consequences of this controlling modern approach have caused some of the recent shifts in water supply practices towards the more inclusive hydrosocial model. Environmental concerns – usually prompted by drought and accompanied by real fears about water shortage for humans, as well as ecology – have lain at the heart of attempts to start to re-engage people with their water use.

Hence, as well as contributing to arguments about the past development of technocratic water governance, Chapter 3 has also provided evidence of the beginning of a hydrosocial approach to water supply across the Global North. This evidence includes the use of an increasingly broad variety of water sources, development of dual pipe systems, and hybrid local and centralised supply systems, as well as demand management through public communications and engagement campaigns. While some of these measures might be seen as largely technical, others are centrally concerned with reconnecting people with sources of water, and also thinking about the consequences of their water use.

In many respects, these changes are exciting, and may, indeed, mark the shift towards a new era of water policy-making. Nevertheless, it has to be stressed that such changes are the exception. Water provision in the London example, as in most of the industrialised world, continues to rely on accessing new conventional water sources, and looks first to technical solutions to problems rather than embracing changes to the hydrosocial contract. Moreover, as will be discussed in more detail in Chapter 4, our homes, habits and expectations are so shaped by modern-era norms of water supply that achieving shifts in water use practices constitutes a significant governance challenge.

4

WATER IN THE HOME

Learning from the past

Introduction

For many of us, our most frequent and habitual associations with water occur in our homes. This chapter aims to explore how we relate to water in the home and how this has changed over time. The rationale is that, by understanding the drivers for change and the extent, style and consequences of change in the past, a more realistic framework for understanding potential changes in the use of water in the long- and short-term future can be developed.

The social science tool that is used to explore water use in the home is the idea of social practices. The key argument here is that we do not want water services for their own sake, but rather for the practices that they facilitate and the benefits consequently derived. For example, water supply allows showering that gives the benefit of feeling clean. Some of the benefits that we now associate with water use include the cleaning of selves, cleaning of food, clothes and home, hydration and waste disposal. Understanding the changing demand for these benefits, and also the changing cultural practices through which they are derived, is part of the process of understanding the changing demand for water in the home.

Chapter 2 has provided the wider basis for changing water resource availability and Chapter 3 has charted how water supplies to cities have developed. Until very recently, it was emphasised, those supplying water perceived their role as satisfying whatever demands existed. In this context, Chapter 4 aims to analyse the causes of some of those demands for water and, specifically, the demands deriving from domestic use.

The chapter begins by examining practice theory and how it can help us to understand water use. The main part of the chapter is a historical narrative about the changing nature of water use practices in homes in England. I have chosen to focus on homes in England because of the ease of access to source material. Moreover, though the discussion covers the full range of water services, particular importance is given to water supply and waste water removal services, which impinge more than surface water drainage on our daily lives at home.

Thinking about water in the home through social practice theory

Water services, benefits and practices

Most urban homes in the industrialised world enjoy the direct benefits of three water services: water supply, foul drainage and surface water drainage. However, we don't want these services for their own sake, but rather for the benefits which they enable us to enjoy, namely hydration; cleaning and preparation of food and food equipment; cleaning of selves; cleaning of clothes, home and other possessions; green lawns and healthy plants; disposal of human waste; and dry home, garden and possessions. The means of enjoying these benefits, and also their extent and importance, vary between cultures and have varied through time.

We can understand these variations in the way these benefits are enjoyed as relating to different water practices. Practices are routinised sets of linked understandings, procedures and engagements (Warde, 2005) that are shared. For example, we have shared understandings of what is meant when the term 'gardening' is mentioned; we also have ideas about why gardening might be done (e.g. to be outside, to grow things, or to enjoy the benefit of having grown things), of the procedures involved (e.g. digging, hoeing, weeding etc.) and the engagements with material objects that are likely to be involved (e.g. with soil, tools and plants).

Different practices associated with one particular benefit may occur with varying frequencies, requiring different skills, different equipment and with different meanings. For example, in medieval England the basic benefit of hydration was often achieved through the practice of brewing and drinking 'small beer' (Mortimer, 2008); in contrast, in a contemporary UK context a significant amount of hydration is achieved through brewing and drinking tea. (It is noteworthy that small beer was only slightly alcoholic, and was used for hydration in locations where the drinking water was not reliably clean.) Table 4.1 compares the two practices. The comparison is useful because the strangeness to us of the practice of brewing small beer can help us to demonstrate how unusual and surprising our contemporary tea-brewing practices would be for the medieval beer-brewer, as well as to others (including our descendants) undertaking different hydration practices in different times/places.

Of course, as well as varying between cultures, there are also individual variations in practices within a culture. You may have a particular way of liking your tea, and our medieval woman may have had some special herbs that formed part of her small beer recipe. The culture sets a broad framework of meanings and the main elements of skill and equipment, but individuals will develop and reproduce these in particular ways to suit their needs and understandings. The cultural meanings of brewing (small beer/tea) may relate to refreshment, but for you and the medieval woman this meaning is supplemented by personal associations – for example, you may associate tea with toast or

TABLE 4.1 Comparison of different practices both achieving the benefit of hydration

Hydration practices	*Brewing small beer in England*	*Brewing tea in England*
Period	Medieval	Contemporary
Locations	In home and purchased from others, or received as part of payment for agricultural labour	In home, at work and purchased from others
Frequencies of brewing	Regular but not very frequent (perhaps monthly)	On each occasion consumed, often on several occasions in one day
Skills and knowledge	• Trust in water and other materials as sufficiently good for brewing • Quantities of which materials, what processes and what machinery (by brewing beer, some potentially contaminated water could be rendered safe)	• Trust in water, milk, tea sources and machinery • Knowledge of temperature of water and length of time before tea bag removal. Knowledge to dispose of (not reuse) tea bag
Equipment	Brewing materials – water, hops, grain, herbs Specialist brewing tools such as barrels	Brewing materials – water, tea bags, milk Specialist tools such as kettles, teapots and mugs
Meaning (example)	Associated with good housewife and good food	Associated with waking up and becoming alert

other specific foods. Likewise, individuals will have particular preferred equipment (mug/drinking vessel) and skills (for example, judging quantities of milk/herbs).

The location of practices

In examining how water practices in the home have changed over time we need to bear in mind how many practices occur variously outside and inside the home. For example, we can use public toilets/baths/water fountains/laundrettes or we can use the equivalents in our homes. We can also eat out or enjoy pre-prepared food rather than washing and preparing food for ourselves. In these respects telling the story of water use in the home involves examining the ebb and flow of water-using (and waste water-producing) activities from the home. Table 4.2 provides a demonstration of how water service benefits can be enjoyed at home or outside of the home.

The location of water use practices is important for those providing and developing water services. If all water services were consumed outside the home then the extent of water infrastructure could be much more restricted; effectively just the provision of services to specific water use 'hubs'. Efficient supply, treatment and reuse facilities could also be concentrated at those hubs. Our current pattern of water service management appears to reach towards the other extreme; water services are provided everywhere. If this highly distributed pattern of consumption is accompanied by centralised distribution systems it means we need a lot of infrastructure. As the

TABLE 4.2 Practices for obtaining water service benefits at home and out of the home

Water service benefit	*Practices using water services at home (Equivalent practice without water)*	*Practices using water services out of the home*
Hydration	• Home tap • Home-brewed tea or beer	• Public spring, well or water fountain • Purchased liquids
Cleaning and preparation of food	Washing food and water use in cooking processes	• Buying pre-washed/ pre-prepared items • Eating out
Cleaning of selves	• Basin wash for hands or feet • Showering • Bathing • Flannel wash • Saunas or similar • Clean clothes	• Outdoor bathing • Communal bathhouses or saunas • Bathing at gym or workplace
Cleaning of clothes	• [Brushing] • Home laundering by hand/ machine	• Use of laundrette, laundry service or dry cleaners
Cleaning of homes	Scrubbing/mopping floors, washing windows, cleaning bathrooms	Living in hotel (or serviced apartments)
Green lawns and plants	Garden and/or indoor plants	Use of parks or other open space
Disposal of human waste	Toilet/[composting toilet]	Public toilet/sheltered outdoor space
Protection from inundation	Location of home, use of resilient building techniques (e.g. stilts), careful placement of precious goods at higher levels	Flood insurance, flood defences and making space for water (see Chapter 7)

historical account below will demonstrate, however, the direction of causality between the centralised/decentralised nature of service provision and the location of consumption is not simple. Where water use practices occur, they both are influenced by and influence the location and nature of water use services required.

Who benefits?

We may think of water service benefits in the home as largely accruing to the individual or family; after all, if you have a shower, you have pleasurable sensation of the hot water falling on you, and it is largely you who afterwards benefits from the enjoyment of your clean body. However, there are also benefits that society enjoys from your 'choice' to wash. If you did not wash you would be extremely aware of the possibility that you might smell to those you meet, and some of those people might distance themselves from you because of your odour; more broadly, widespread personal hygiene has benefits for society in terms of minimising pest and disease transmission. You therefore perceive the need to wash to be required not just for your private benefits but because being clean is something that society requires of you. Likewise, society benefits from you having clean food and hence being able to labour, or not demanding sickness services, or not constantly feeling anxious about whether your home is going to be flooded. In these respects, water services provide both individual and collective benefits.

The various individual and collective benefits from water-using practices change with our perception of what is important and why it is done, in other words, with the meaning of the practice. As the following discussion will show, during the early modern period in Europe controlling body odour through washing was not perceived as important (Ashenburg, 2009: 78). Whether we are emphasising individual or collective benefits from water practices matters because it influences the way we think it is appropriate to pay for water services. If our understanding emphasises the communal benefits of people washing associated with odourless-ness, public hygiene and health, then we would be happy for it to be paid for collectively. If our understanding stresses the individual benefits of such services, then our approach would put more emphasis on individuals paying for the benefits they enjoy.

In summary, we can call the rationale or benefit achieved from undertaking a water practice its 'meaning'. Practices vary between and within cultures in relation to their meaning, but also in relation to their location, frequency and associated skills and equipment. The meanings we seek from water practices include both individual and societal benefits.

Changing water practices in the home

In exploring past water practices in the home, this section focuses on England from the medieval period to the present day. As elsewhere in this volume, the historical information serves as an exemplar of how the practices that we currently relate to water have changed. Rather than covering all water practices, there is a focus on making the body clean, with some links to disposal of bodily waste. I have focused on these rather than hydration, for example, because they have had more profound impacts on the flows of water in and out of the house. The account is interesting because it indicates some very significant changes in habits and understandings.

In focusing on historic England, the concern is with a country with the legacy of a Christian culture. This has important implications. While most world religions lay out specific ritual bathing techniques, this is not the case with Christianity. This difference between the religions may explain why habits of bathing and cleansing may have been far more variable in Europe and other Christian-oriented cultures than in Asia or other areas where, for example, Islam or Hinduism specify washing practices to be followed (Ashenburg, 2009).

Medieval England (1300–1500)

Some medieval households had good access to water with a spring or well in a yard or immediately adjacent to their homes; others would need to carry water some distance from rivers or conduits, as well as collecting it from roofs or paved areas. Given the work needed to get water into homes, it is likely to have been used sparingly, and to have been reused for progressively dirtier uses. Like in many emerging economics today, quantities of waste water produced might therefore be relatively small, and disposed of in designated locations. Moreover, as Skelton points out, in practice household waste water quantities were small compared to the waste water produced by urban agriculture and (animal) transport (Skelton, 2016: 21). Many significant uses of water, for example, supplying water to animals or laundering, were achieved by taking the job to the water rather than the other way round. Nevertheless, even this had to be done with care, as the water source had to be protected from contamination in order that others could use it.

Washing of hands and faces was typically carried out in mornings and evenings, while the wealthy also washed hands before and after every meal, and feet were soaked typically after a journey (Mortimer, 2008: 196). Heating water to bathe the whole body required large amounts of water and energy, and was therefore selectively done in poorer private homes, but occurred more frequently among the wealthier, including merchants and the aristocracy. Dorothy Hartley writes of both mobile and static wooden tubs, the latter 'arranged under a hook in the ceiling from which hung a tent-like canopy', keeping out draughts and affording privacy (Hartley, 1964: 174). Cold washing in rivers was available to everyone, and continued to be widely used far beyond this period. For example, a Captain Eyre writes about three night-time excursions to bath himself in the River Don in Sheffield in 1647 (Eyre, 1875: 48, 50–1, 57). In practice, outdoor surface water bathing is likely to have been regularly used by people employed in particularly dirty professions, and other healthy people might have enjoyed such bathing in the summer. Nevertheless, an unwashed smell was seen as an indication of virility, particularly in rural areas (Mortimer, 2008: 196–7).

'Urban water management', as we currently understand it, was significantly limited for the basic reason that relatively few people lived in towns; in 1377 there were approximately 170,000 urban dwellers representing 6 per cent of the UK population (ibid.: 11). Nevertheless, many people accessed towns on a regular basis for the purpose of exchanging goods. The need for access to collective water services outside of the home was particularly true in urban areas where, as was discussed in Chapter 3, water resources came under pressure from both domestic and industrial use, as well as pollution. In Exeter, for example, there were shared sources of water, communal toilets on the town bridge and a specified location for human waste to be disposed of outside the town (ibid.: 7–13). Collective all-body hot water washing was usually available in urban areas through bathhouses or 'stews', reintroduced to Northern Europe by crusaders returning from Turkey in the eleventh century. Baths were heated, often linking with adjacent sources of heat such as bakeries, and boys would run through the streets shouting when they were ready (Worsley, 2011: 109). People would enjoy a soak in an individual tub scented with herbs in a communal room. In 1378, most of the 18 bathhouses (or 'stews') in London were located in Southwark, where they were seen as a means of recreation, and also associated with prostitution.

Medieval toileting was not a particularly water-oriented activity. Buckets served most homes and were emptied in some communally designated location, such as a dung heap, or at downstream locations on a river. Large houses or castles had garderobes or stool rooms to provide toilet facilities for the gentry, some with disposal systems built in (to moats, for example), and some requiring emptying by servants. In stark contrast to today's norms, the King was always accompanied when visiting the stool room by the Groom of the Stool, a very important member of the royal court, as he was always extremely close to the monarch (ibid.: 156). Members of the household who were not part of the gentry would use collective toilets known as 'commonjakes' – effectively a row of adjacent seats. In urban areas, similar communal public toilets were provided where waste could be removed with ease – for example, over rivers and streams. There were 13 public toilets recorded in medieval London, including one with 84 seats (ibid.: 154).

The early modern period (1500–1800)

The shift from medieval to Tudor England coincided with some significant changes in water use in relation to personal cleanliness. Public bathhouses had long been seen as linked to the spread of plague. Warm water was understood to leave the pores of the skin dangerously open to substances that could upset the balance of the body's humours, and expose people to diseases like syphilis and women to pregnancy. In contrast, pores blocked by dirt sealed the body off from infection and were thought to ensure good health. This image of bathing creating vulnerable exposed bodies circulated and recirculated whenever there was epidemic disease in Europe from the Black Death in the mid-1300s onwards. Bathhouses were vulnerable to such accusations because they were already morally suspect; this arose both from their link with prostitution and licentiousness, and their association with eastern habits and religions (Ashenburg, 2009: 111). The latter was especially present in people's minds because of the recent reconquest of Spain from the Moors in 1492, during which many bathhouses were destroyed. The increasing challenge of providing sufficient water and fuel in urban areas may have been an additional factor contributing to bathhouses' demise (Worsley, 2011: 113).

France led the upsurge against washing by temporarily closing bathhouses whenever there was illness through the early 1500s, and permanently closing them in 1538. In Britain, Henry VIII banned the stews in 1546, and in 1576 Spain closed their remaining bathhouses. Though the bans only applied to public bathhouses, they are seen by many to mark a broader shift in social norms. Ashenburg quotes a French commentator at the end of the 1500s voicing a minority and old-fashioned regret about 'having lost' the custom of bathing (Ashenburg, 2009: 100). The strength of moral judgement associated with this change is indicated by the records from the Spanish Inquisition; converts from Islam or Judaism were expected not to bathe, or the sincerity of their new faith would be questioned (ibid.: 111). Elizabeth I is quoted as having been unusual in having a bath once a month 'whether she needed it or not' (Bryson, 2010: 99). Drawing on British and French travellers' surprised commentaries on trips to Turkey and the East during the 1500s and 1600s, Ashenburg concludes that, for Europeans in this era, 'a culture where people bathed several times a week and regularly washed their genitals was exotic and even bizarre' (Ashenburg, 2009: 105).

It is important to understand that those who became shy of water continued to emphasise personal cleanliness, but their means of achieving this changed from bathing to clean linen. Linen was viewed as a way to soak up sweat and to achieve cleanliness that was more scientific and advanced than processes of bathing, which were seen as out-dated. The emphasis on linen is highlighted in portraits of the period that show fashions shifting to reveal more linen; a sign of conspicuous consumption for those that could afford the servants to launder it (ibid.: 110; Worsley, 2011: 114). In terms of water use in the home, the impact of this change for the wealthy is a shift from the use of baths, and the employment of servants to prepare them, to a larger quantity and high standards expected in relation to laundry. The latter continued to be carried out in external public locations by the poorer working people, and to use soap as well as urine in the cleansing formulae. In terms of the cleanliness of the less well off, those in urban areas ceased to have access to bathhouses. For the rural poor the change may not have had so much significance; apart from a cold-water dip in a river or stream, becoming bodily 'clean' remained a luxury for the urbanised and the rich.

Although some shift away from bathing for the vast majority was propelled or at least supported by medical advice at this time, the same medics also advocated bathing for those who were ill – for example, to address Henry VIII's leg ulcer (Worsley, 2011: 117). In this way, for a period, bathing in warm water came to be associated purely with illness and therapy. This medicalised perspective on bathing appears to have enabled two different developments that served to bring bathing back into prominence among the English gentry in the 1700s. First, spas such as Bath, Buxton and Tunbridge Wells emerged as centres for curative water treatments during the seventeenth century, and then came into their own as venues for high society gatherings during the eighteenth century. By the end of the eighteenth century, therefore, the town of Bath was featuring prominently in Jane Austen's novels of *Northanger Abbey* and *Persuasion* as a centre for the gentry to socialise. Spa treatments involved various combinations of immersion and drinking of mineral waters. Meanwhile, from the late 1600s doctors and other commentators began to promote cold-water bathing in non-mineral settings for invigoration and hence for maintaining health. Cold-water immersion was seen as both cleansing and strengthening (Ashenburg, 2009: 129; Worsley, 2011: 118).

Over and above changing cultural understandings of cleanliness, the management of water in the home shifted in this period due to the processes of urbanisation England was experiencing. Whereas only one in 12 English people grew up in a town in the 1370s, by the 1600s it was one in four (Mortimer, 2013: 24). As discussed in Chapter 3, town authorities took some responsibility for maintaining the quality of water sources, including rivers and streams. Many towns also employed labourers to maintain the drains that served the main streets (Skelton, 2016: 84). Beyond the responsibility of town authorities, however, individual residents had responsibilities for town ditches and drains located in the streets outside their homes. Skelton (ibid.: 88) describes the case of a certain Elleanor Harris of Whitehaven, who sought legal support for her right to clean the sewer outside her house, which was being hindered and stopped by a neighbour. She argues that: 'Elleanor represents the majority of urban residents in the period, people who completed their own sanitation duties day after day and did not cause problems for either their urban governors or their neighbours.' Nevertheless, there are examples of people flouting sanitary expectations, for example, Mortimer cites examples from the village of Ingatestone in Essex, in the 1560s, where people are prosecuted for having privies over the common sewer, and also for leaving dead carcasses on the public highway (Mortimer, 2013: 34).

The late 1590s saw the invention of the water closet by John Harrington, and the installation of one of its number at the Queen's castle at Richmond, as well as in Harrington's own house (Skelton, 2016: 33). For most people, however, the bucket or external earth closet remained the main toilet facility. In richer houses, chamber pots emptied by servants achieved the same function. Foul waste was kept in cesspits beneath or behind the houses. When cesspits became full, house-owners/landlords could pay for them to be emptied by night soil men and sold to farmers as essential agricultural fertiliser. The issue of cesspits overflowing and/or contaminating groundwater is likely to have been a perennial problem. Samuel Pepys, for example, refers to the turds in his cellar in a description of when his neighbour's cesspit overflowed in 1663 (Bryson, 2010: 503).

The return of water (1750–)

In the wake of changing ideas about cleanliness, shifts in fashions, medical understandings and urban life occurred. During the period when bathing was frowned upon, complex constructed costumes with wigs, corsets, lace and heavy scents for both genders hid much of the body. In the late 1700s, these gave way to more simple dressing that exposed the natural hair and skin, with lighter floral scents now confined to women (Ashenburg, 2009: 145). Whereas the previous fashions had not required limited bodily cleanliness beyond the face and hands, new perspectives gave an emphasis to cleanliness all over and, specifically, the need not to smell. Second, a new understanding of sweat emerged, emphasising the need to clear pores to let sweat and dangerous internal substances out of the body (rather than insulating it from dangerous external substances, as previously understood). Bad smells or 'miasmas', including bad smells from people, were increasingly seen as a vector of disease (Worsley, 2011: 118). Third, English cities began to grow, both enabling and driving these changes. Noise, smoke, industrial emissions, overcrowding and lack of sanitary facilities made the poor and manufacturing parts of cities increasingly unpleasant. Better roads let middle-class districts spread up-wind and up hills, where fresher air and, increasingly, piped water was available (ibid.: 120; Hey, 2010: 113). Poorer areas tended to remain where there was industry, continuing to rely on water from the same surface and groundwater reserves as previously.

The impact of these changes for the middle and upper classes was that washing became more common again, although at neither the frequency nor the intensity that we might expect today. Ashenburg likens the use of baths and bidets in Europe in the late 1700s to the use of jacuzzis and saunas in the USA today: 'the important thing was that they existed, and a growing body of opinion valued them' (Ashenburg, 2009: 149). One of the issues with bathing was exposure of nakedness, particularly for women. Consequently, many women bathed wearing bathing dresses, and extensive use was made of 'slipper baths' (so-called because they were shaped like a large slipper and hence enveloped the body). Writing to his son in 1749, Lord Chesterfield recommended frequent washing and use of a 'flesh brush', for the sake of both his own health and in order not to be offensive to others (in smell). Immersion, particularly warm water immersion, still aroused some suspicion, but the use of warm water to perform a sponge wash became the norm. In one account of an English country house in the 1860s, warm water was provided to bedrooms three times a day (ibid.: 165–6).

The continuing problem of contamination from cesspits was much augmented with the spread of the flush toilet from the 1780s. Though invented by John Harrington in the 1590s, the flush toilet did not become popular until the provision of piped water eased the logistical challenges of fitting it, and toilets became popular with middle-class households in London. Their impact was to massively increase the liquid content of human waste from households, causing cesspits to overflow far more frequently. Due to this growing problem, by 1815 it became permissible to drain refuse (that is, foul sewage) into a sewer in London (Allen, 2008: 29), and by 1848 it became illegal not to connect new properties to a sewer.

The coming of the flush toilet had several important implications. First, the shift towards using water to carry waste meant that toilets could more realistically be moved inside the house than previously. This shift took over a century to occur, but it was the flush toilet that made it possible. Second, and consequently, the home acquired a new form of connectivity to other homes and to the places where sewage was disposed. It is perhaps not surprising that this latter element was one of the reasons why some people resisted the development of sewers. Whereas a water supply network evoked connection to the countryside, the sewer network highlighted a household's connectivity to other houses, and to unsavoury locations where sewage was disposed of. As such, some perceived sewers as a violation of privacy and domestic autonomy (ibid.: 39). There was a specific concern that disease could be carried through 'sewer gas' moving (in complainants' imagination) from the poorer districts into the more salubrious parts of town. The latter concerns were given particular prominence when the sewer gas was attributed as the cause of the Prince of Wales contracting typhoid fever in 1871 (ibid.: 44).

During the late nineteenth and early twentieth centuries, the provision of warm water by servants began to change for middle-class households, with the spread of the bathroom. The standard formula at this time was a Spartan white-tiled room: 'the bathroom's script was clear, cleaning was a job to be done and this was the place to do it' (Shove, 2003: 101). Cleanliness – measured by smell as well as appearance – became seen as a mark of respectability. In *Road to Wigan Pier*, George Orwell comments that, when he was a child (i.e. during the 1900s), he was taught to understand that the working class smelt. A sharp class association with cleanliness is also communicated by Shaw's play *Pygmalion*, first performed in 1912, in which the working-class Eliza Doolittle is reluctantly forced to have a bath as the first step in 'educating' her so that she can be passed off as a duchess (Ashenburg, 2009: 231). Shaw can be seen as promoting middle-class understanding of limited working-class washing facilities when he ascribed the following comments to Eliza Doolittle: 'I tell you, it's easy to clean up here. Hot and cold water on tap, just as much as you like there is... Now I know why ladies is so clean.'

Some attempts to enable working classes to have access to washing facilities occurred from the nineteenth century, with the provision of communal wash-houses such as those in the model village Saltaire, near Bradford. These, however, did not prove popular for washing (of self), though they were extensively used for laundering clothes (Ashenburg, 2009). In 1918, legislation was passed requiring that bathrooms were provided in all new houses, not just those of the middle classes (Worsley, 2011: 129). However, there were many pre-existing homes that continued not to have these facilities well into the 1950s. Fears of smelling continue to be reinforced in adverts for new soap and deodorant products, developed in the 1920s and 1930s. Looking back on a working-class childhood in 1930s London from a middle-class retirement, one commentator tells of the ritual of a weekly bath for three children sharing a galvanised tub of warm water in front of the living room fire (Ashenburg, 2009: 237). By the late 1930s the majority of homes in Britain had running hot water, though many still did not (ibid.: 260–1). These developments indicate the path of washing from a mark of distinction between classes to a question of social acceptance for everyone.

In the inter-war period, houses were built with separate toilets and bathrooms. After the Second World War, toilets and bathing facilities were increasingly integrated into one room. The aesthetics of bathrooms also shifted. Rather than the clinical cleanliness of the early twentieth century, coloured tiles and bathroom suites developed alongside a proliferation of toileting and bathing products. For a long period, the bath rather than the shower dominated homes in England. Lack of showers in mid-twentieth-century England has been related to the lack of central heating and poor insulation in many homes, and also to the relatively low water pressure compared to Continental Europe. From the 1970s onwards, showers were increasingly provided as well as baths. An increase in the prevalence of central heating and the development of new shower technologies has coincided with the increasing dominance of the shower in the English patterns of bathing. This has not necessarily reduced water use, as one or two showers a day has replaced a bath every other day (Southerton *et al.*, 2004). Bathing is variously associated with an invigorating wake-up or a relaxing de-stress. Bathrooms have become bigger, and increasingly houses have en suite bathrooms for at least some of the bedrooms.

The twentieth century has witnessed periodic concerns about the quantity of water used by the public, arising when droughts occur (Taylor *et al.*, 2009). Through limiting water availability to some, droughts force water utilities and regulators to ascribe legitimacy to some uses and illegitimacy to others. The perception of legitimate uses has shifted through time. For example, during the 1976 drought, public supplies were restricted but supplies to industry were prioritised. In contrast, 1995 saw industry restricted in favour of the public. Towards the end of the twentieth century and in the 2005–6 drought in the twenty-first century, there has been increasing recognition of the inevitable episodic nature of droughts, and hence of the need for more systematic interventions to save water.

Conclusion: learning from the past

The account of bathing is captivating because it involves dramatic changes in practices. Specifically, the shift from a world in which bathing was widely practiced, at least by the middle and upper classes, to one in which bathing was almost completely eschewed is a bizarre one. Our surprise undermines any assumption that history involves linear progress from some uncivilised past to the present. Instead, it suggests a 'jerky' process of history as, first, there was bathing and then there was not, and now there is again. Breaking down the linear view of history as 'progress' (in bathing and many other things) is one insight that can result from this account. But it is useful to pause a little longer to consider what else we can learn from the changing use of water in the home.

First, the account exposes how individual practices need to be understood within a societal context. We may have considered our personal bathing habits a matter of individual choice – which, of course, to some extent they are – but the history of bathing reminds us that our own choices are highly constrained by technologies, systems and societal expectations. As Shove describes it in her study of cleanliness: 'private habits are constructed as people steer their own course through culturally and temporally specific landscapes of legitimating discourses' (Shove, 2003: 94). In other words, water-using practices have meanings and these meanings are made and remade collectively.

Second, the impact of positioning individual choices in a societal context is that it undoes assumptions we might make about the 'value' of a particular set of practices. It highlights that any idea of our current bathing practices as 'superior' to those of other eras needs to be understood as a typical, but rather unenlightened contemporary view; of course, we think it is superior because it is what we have been taught/constrained to do. Informed by history, we can notice that what people think is the superior practice has changed in the past, and is likely to change again.

Third, as well as being entangled in societal meanings, the account highlights how much water practices are enmeshed in collective physical systems of water supply and disposal. In the past, many of these systems were small collective systems occurring outside the home. During the period studied, we can see that most of our bathing and toileting practices have moved into the home and have also become more intimate and private. Meanwhile, the systems supporting these practices have spread more widely, connecting broader geographical areas with demand for water and/or flows of waste water. In this respect, bathing and toileting has become both more private and more public.

Finally, the historical account raises questions about how practices change. It is clear that there is no simple explanation for changes in water practices, but rather a combination of factors at work including technology, systems of provision, medical knowledge, popular understandings of bathing and toileting, and edicts from authority:

- The existence of a technology such as the water closet clearly has a role in changes occurring; but it is equally clear that the technology is not the decisive factor – the toilet technology existed for 200 years before it became popular.
- The combined existence of technology and the systems (water supply, sewage) to support it could be argued to be important. However, many toilets were fitted before sewers became the norm – so, once again, these are not the only factors.
- It is also clear that science and medical understandings have played a role in contributing to popular understandings of what is attractive and healthy; however, there is no simple causality between them. For example, the case of medieval bathhouses illustrates how medical understandings interacted with other suspicions about their foreignness and licentiousness. Rather, it seems that medicine offers legitimating discourses through which changes that may be arising for another reason can be supported and are further promoted (ibid.: 95).
- Additionally, bathing has close links to the way we present ourselves to society. In this way, how we smell, what we wear and how our hair looks all offer signals through which we can assert a particular desired identity, and through which other people will judge us. Identity issues about being different (e.g. cleaner, having new widgets in homes) and/or the same as others (e.g. not smelling, having the same widgets in homes) appear to have played an important role in driving some changes forward, such as the increase in bathing over the latter part of the twentieth century. For Shove, medical knowledge, perceptions of self in society and the role of pleasure in bathing can all be understood as part of the changing understandings of what bathing is for (its 'meaning') and relate to 'core concerns to do with self and society, body and nature, pleasure and duty' (ibid.: 114). She suggests that these meanings evolve such that 'changing practices represent new ways of 'fixing' or resolving a handful of persistent preoccupations' (ibid.: 95) within the context of technologies and systems available.
- Finally, interventions by authorities to promote changing bathing practices have not always been successful when first introduced, but have shown a pattern of achieving their aims in the long run. In this respect an edict by an authority organisation both appears to capture and promote changes that are underway in society anyway. Hence, while the closure of bathhouses in the 1500s had a profound effect, the introduction of public bathing facilities in the 1800s did not have the same immediate impact, but working people have acquired the habits of hygiene that the upper classes desired for them, and more, during the twentieth century.

Overall, given an aspiration to encourage more sustainable patterns of water use, surface water management and waste water disposal, reviewing past washing habits reveals some opportunities, but also some cautionary warnings. In this respect it throws light on the nature of a hydrosocial model of water management. In terms of opportunities, if habits and meanings of water uses have changed in the past, it shows that they could change just as dramatically and unexpectedly in the future. In particular, given that people have previously switched from 'bathing' to 'not bathing', it suggests that unimaginable as well as imaginable change can happen. It is also apparent that interventions by authorities to achieve changes in water practices can and have been successful. Finally, we can see that there are a wide range of factors that influence water practices including technologies, systems of provisions and the complex realm of water-practice meanings relating to self and society, body and nature and pleasure and duty. Those seeking to change water practices can fruitfully think about these different factors as providing a variety of means of potential influence. In terms of cautionary points, it is clear that, since the spread of piped water, there has been a dynamic that has worked to increase water use, while growing population and changing patterns of waste disposal have resulted in the long-term deterioration of watercourses. While it would be inappropriate to assume that these dynamics associated with technocratic water management will continue forever, making purposeful interventions to change these long-established patterns is likely to be challenging.

In terms of the arguments of this book, the primary contribution of Chapter 4 has been to provide some of the detailed domestic processes that have both contributed to and resulted from the spread of technocratic water management. Additionally, however, the discussion has also introduced the 'practice' perspective that will be crucial in thinking about contemporary water use practices. Informed by this historical perspective on changing practices, Chapter 5 turns to consider the way that contemporary water practices are understood and how changes to practices are promoted.

5

UNDERSTANDING WATER PRACTICES AND MOBILISING CHANGE

Introduction

It is already clear from Chapter 3 that forward-looking utilities have shifted their perspective from simply meeting all water demands unquestioningly to embracing new ideas about managing water demand. We also know from Chapter 4 that water practices have shifted significantly in the past and hence, perhaps, that they might shift in the future. The aim of Chapter 5 is, therefore, to explore the strategies for intervention that might be made to help 'desirable' changes in water practices to happen.

In terms of the hydrosocial governance cycle depicted in Figure 2.5, this chapter focuses on the arrow leading from the Governance box to the People box. Representing the process I have called 'mobilisation' (also sometimes referred to as 'influencing' below), this arrow depicts how those acting with some collective interest are seeking to influence public water use practices. Mobilisation contrasts with 'participation', depicted by the arrow in the reverse direction from people to governance (and discussed in Chapter 9), which concerns the way that the public are seeking to (or sought to) influence collective action. In the wider context of the book, the elaboration of mobilisation and participation is important because these processes have a much-enhanced role in the context of hydrosocial water management as compared to a technocratic approach. This chapter explains what has been done in moving towards greater mobilisation, as well as suggesting how mobilisation might be done better.

There are a variety of water practices that practitioners may try to influence. Many of these relate to water consumption, but they may also relate to the use of waste water systems and to the drainage from households' roofs and drives (see Box 5.1). For the sake of brevity, and also because of the position of this chapter within the narrative structure of this book, the following discussion concentrates only on the attempts to understand and mobilise change in household water consumption. It should also be noted that, though the focus is on water use in the domestic household, many of the

BOX 5.1 AREAS WHERE PUBLIC ACTION IS TAKEN OR SOUGHT IN RELATION TO WATER SERVICES

- reductions in the quantity of water used, either in general or at times of drought; also reductions in the quantity of hot water used as a means of impacting upon energy and carbon emissions
- development or acceptance of alternative 'fit-for-purpose' local supplies of rainwater, grey or recycled water on either a household, neighbourhood or system-wide basis
- reductions in the peak-time demand on surface water sewers through storage or infiltration of rainwater falling on roofs and drives, either by the household (for example, in roof tank or in a garden) or on a street/neighbourhood level
- changes in the use of foul sewers – for example, by preventing the disposal of sanitary products and waste grease through sewers, promoting use of the solid waste stream instead
- increases in the awareness of and preparedness for floods – for example through signing up for local flood warnings, through having a household flood plan, through investing in flood-resilient fixtures and fittings, and through supporting neighbours and/or the wider community as a voluntary flood wardens
- public involvement in the monitoring and maintenance of publically owned water and green infrastructure – for example, through clearing (or alerting authorities about) blocked gullies, through participating in litter or vegetation clearances and through collating and providing ecological information

points made are equally applicable to other non-organised users of water services such as small and medium-sized enterprises.

This chapter looks both at water utilities' understanding of water activities and the means through which they are trying to promote changes in those activities. As the combination of 'concepts' and 'organisations' within the Governance box in hydrosocial governance cycle (Figure 2.5) might lead you to expect, understandings influence the options considered for mobilisation. I'll begin by discussing conventional means of understanding water 'behaviours' and the ways that these understandings lead to specific approaches to influencing people. I will then develop some critiques of these approaches. The second section moves on to discuss some emerging ways of understanding water 'practices' and alternative means of mobilising changes in those water practices. The chapter largely draws on evidence from attempts to influence domestic water use practices in England, a country in which 40 per cent of the public water supply is used in households.

The overall story is one of increasing recognition of the complexity of everyday life and growing sophistication in considering whether and how it might be influenced. As will become apparent, understanding and influencing everyday practices may occur through direct influence on a householder; equally, it may occur through influencing the organisations and systems that shape household practices, such as household equipment manufacturers, plumbers, builders, food-sellers and the makers and advertisers of household and personal cleaning products. One of the shifts in recent years (partly, but not exclusively linked with practice theory) is that household water use practices are recognised as the dynamic product of the ongoing interaction between all of these stakeholders. This said, it has to be recognised that any organisations seeking to achieve widespread change in everyday life face an uphill battle, and that the motives of those seeking change are likely to be questioned, alongside the legitimacy of any claim they make about a general 'need' for change.

Understanding water use behaviour

The science of understanding water use behaviour emerged from the 'predict and provide' logic described in Chapter 3. For many years, understanding water use behaviour had nothing to do with influencing that demand, just anticipating its rate of increase and hence justifying investment decisions. As environmental concerns have made it more difficult to develop new water sources, there has been a shift towards understanding water use behaviour in order to change it.

The basic metric through which water utilities understand and predict domestic demand is 'per capita consumption'. As an example, the English government's water strategy document, *Future Water*, provides a set of measures of per capita consumption in different European countries, shown in Figure 5.1.

A central component of forecasting is extrapolation – that is, assuming that the trends of the recent past will continue into the future (Sharp *et al.*, 2011b). The science of this process has become more sophisticated with time. Forecasters are increasingly using demographic characteristics to predict how and where demand will change (Butler and Memon, 2006); per capita water consumption is therefore linked with a variety of characteristics including household size, property type, age of household members, income, number of bathrooms, temperature, garden size and the price of water (Jeffrey and Geary, 2006: 308–9).

In addition to overall per capita consumption, another important factor in the forecasters' tool kit is an understanding of how the water that goes into the home is used. Figure 5.2 shows average per capita uses in English and Welsh homes for different components of water use; it demonstrates high water use through toilet flushing and personal washing using baths, sinks and showers. Such information can be studied in historical context too – for example, between 1976 and 2001 the proportion of household water used for personal washing increased, while that used for toilet flushing decreased (Butler and Memon, 2006: 9). These dynamics of how the total demand associated with different 'micro-components' change are complicated as

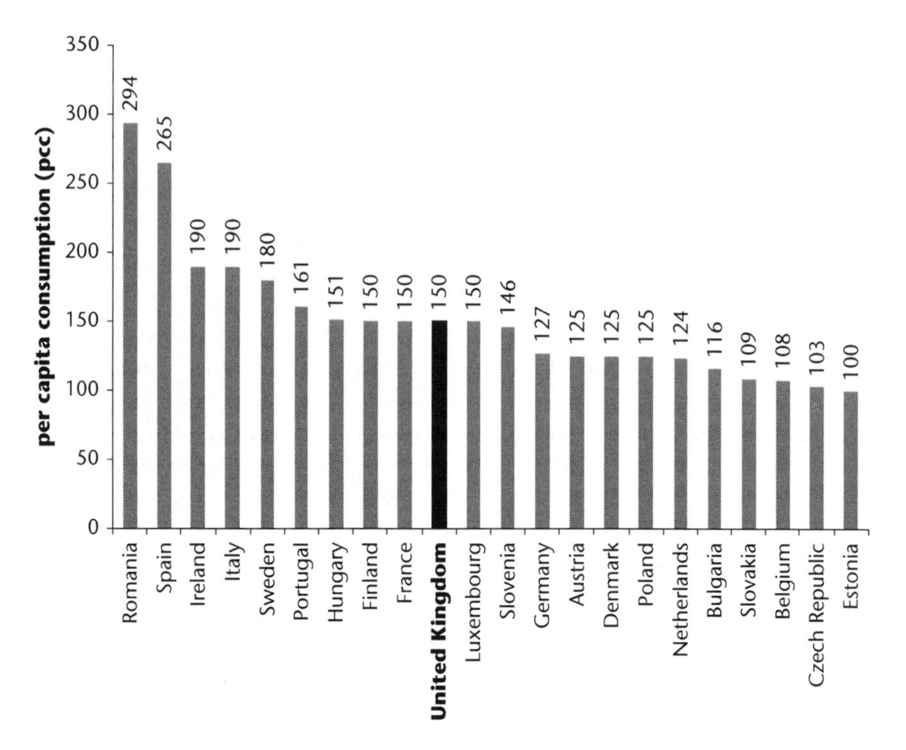

FIGURE 5.1 EU per capita daily water consumption (litres/person/day)
Source: From Defra, 2008: 21, based on Waterwise data, 2006 (no citation), with permission

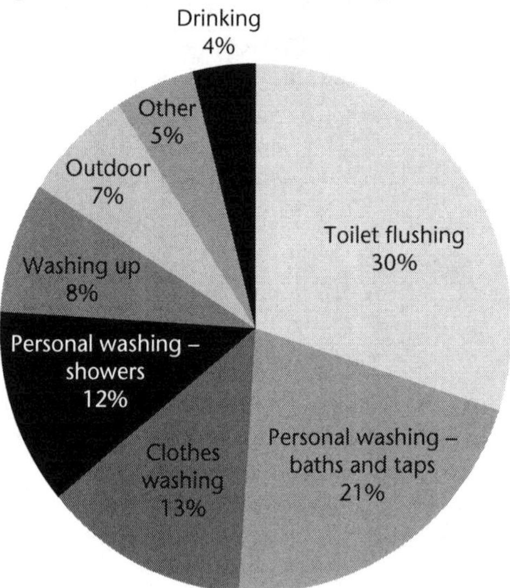

FIGURE 5.2 Average water use in English and Welsh homes, taken from Waterwise fact sheet for consumers
Source: Waterwise, 2013a

they relate to both the changing nature of appliances and changing frequencies of use. For example, although washing machines have become much more efficient, they have also become more widespread and are used much more frequently (Shove, 2003). Figures about how much water is used by different appliances are therefore supplemented with an understanding about their distribution and frequency of use. Expert knowledge is then used to predict how distribution and use will change in the future, and this enables future use to be forecasted.

Forecasts offer the means through which the potential impact of different measures for influencing demand can be evaluated. For example, in their evaluation of means of reducing water demand in the UK, Sim and his colleagues concluded key factors enabling a moderation in demand would be the spread of metered charging and, to a lesser degree, the diffusion of water-efficient devices including washing machines, low-flow taps and low-flow showers (Sim *et al.*, 2007). Under altered political circumstances, they note that the imposition of compulsory metered charging could have a greater impact on demand. (As discussed in Chapter 3, such 'progressive' charging is now underway in London.)

A recent development of 'behaviour'-based studies has focused on understanding and forming generalisations about attitudes to water management and water saving (Opinion Leader, 2006; Defra, 2009, 2010). By investigating and forming generalisations about attitudes to water, these approaches seek to inform forecasting and behaviour change campaigns. An example of such a study was carried out by consultants Icaro for Defra and published in 2010 (Defra, 2010). The study sought to understand household attitudes to the retrofitting of water-efficient devices relating to water conservation, flood risk, sustainable drainage and pollution prevention. Using day-long deliberative forums in selected locations around the UK, a sample of people selected to incorporate a variety of environmental attitudes were asked to discuss and explore different options for retrofitting water-sensitive devices. The deliberative forums were followed up by some home interviews and site visits to locations where water-sensitive retrofitting had taken place. In terms of water conservation, the study concluded that UK residents showed significant willingness to accept water-efficient devices in new homes, including rainwater harvesting and grey water use systems and were willing to consider retrofitting. However, their awareness of such devices was limited, many had doubts about whether and how the devices would work in practice, and the incentives for fitting such devices were not very significant while water remained cheap and was unmetered for many. To address these issues, the report suggested the need for a 'big narrative' about water conservation, accompanied by better operational information.

Influencing water-using behaviour

There are many different actions that have been tried in attempts to influence water-using behaviour. Broadly, these fit into four categories (Jeffrey and Geary, 2006): 'regulations' (absolute restrictions), 'technological change' (achieving changes through different technologies in the home), 'incentives' (the use of finance or other means to entice changes) or 'persuasion' (informing users about the extent of their demand or its impact elsewhere to try to persuade them to change). Clearly, these do not have to be used separately and, as will be seen, there are overlaps in terms of how different instruments operate, but they offer a useful organising device for the discussion below.

Regulations – for example, water restrictions

Regulations are absolute rules prohibiting the use of specific behaviours or technologies. The most extreme or enforced form of regulation is rationing, involving physical restrictions on the time that water can be accessed or the extent that can be obtained. More commonly, regulations impact directly on water users by restricting particular uses – for example, outdoor watering of plants using a hosepipe. In developed countries, at present, the onset of temporary regulations (or 'restrictions') is typically triggered by seasonally low storage in reservoirs.

Economic theory suggests that the advantages of regulations are that they are simple and clear and hence may be relatively easy and cheap to enforce. The disadvantages, however, are that they apply a broad-brush solution across all, without discrimination (Jacobs, 1993).

There are also choices in the way that water restrictions are applied. Whereas the UK approach has focused on the black and white factor of whether households are permitted to use a hosepipe for garden watering, in Australia water utilities typically identify multiple levels of water restrictions involving progressively harsher prohibitions on outdoor uses. An advantage of such a progressive approach is that the relatively minimal prohibitions associated with the lower-level restrictions function as a clear communication tool, highlighting that water is potentially liable to shortage (Sharp, 2006).

The switch from a context in which the public are encouraged to consume as much water as they wish to a context in which certain uses are banned can be experienced very negatively. Effectively, those who usually undertake the prohibited activity are made to feel responsible for the drought. This element of moral judgement inherent in water user regulations means that they have been branded by Sofoulis as 'user-blaming' (Sofoulis, 2011: 807), and it is clear that, at present, such restrictions typically 'punish' those who use water outdoors to care for gardens while leaving other potentially prolific indoor uses of water unrestricted.

Apart from these direct regulations concerning water use, regulations *indirectly* influence the public and their water use in many ways. For example, since the late 1990s UK water companies have been required to promote water efficiency to their customers, while (as discussed below) restrictions on the producers of bathroom products mean that all new toilets purchased since 1999 have drawn on less water in order to flush than was commonly the case previously. Restrictions can also influence the growth of demand through planning controls on the construction of new properties (Jeffrey and Geary, 2006). Because the direct intended influence of these regulations is to change water organisations' persuasive messages (the promotion of water efficiency) or to shift the nature of technology (water fittings regulations) they are discussed in the sections below.

Technological change

Technological change means using technologies to achieve resource efficiencies. As the history of water in the home from Chapter 4 has illustrated, implementing new technologies involves both the invention of new technologies and their subsequent diffusion. Governments can and do support technological development through research funding in relation to water management, as well as many other areas. However, much effort is also directed at supporting the diffusion of existing technologies, which might occur through information provision and/or through regulation.

Currently a variety of technologies exist that could increase water efficiency. These range from efficient variations on currently universal technologies – for example, the low-flow tap – to technologies requiring significant restructuring in terms of household infrastructure, such as grey water recycling. If implemented universally, even the known and trusted technologies could achieve a significant decrease in water use (Sim *et al.*, 2007).

An example of government support for the diffusion of technology in England is the aforementioned *Water Supply (Water Fittings) Regulations* that have reduced the maximum amounts of water used in new washing machines and toilets (Stationery Office, 1999). This example can be seen as a classic case of a 'fit and forget' technology in that the change is organised by government making a rule that manufacturers have to follow. Hence, the public will buy fittings that appear the same and work the same, but each use will draw on less water.

In the absence of an overall rule made by government, other 'fit and forget' technologies require consumer 'choice' for more efficient versions at the moment of appliance purchase. In England at present, for example, householders can choose to purchase shower fittings that are relatively more or less water efficient. Governments can support the diffusion of new technologies to domestic households through rules (like the fittings regulations), but also through subsidising training for relevant professionals such as plumbers or planners, by creating standards for houses/products, and through product labelling. For example, low-flow showerheads featured as one of the environmental measures that developers of new homes could fit in order that their development achieves a 'greener' designation within the 'code for sustainable homes'. Additionally, UK consumers refitting their bathrooms will soon have access to a water label through which to judge the water efficiency of products such as showers (WRAP, 2013). Crucial to the success of such measures will be whether the efficient fitting is perceived to provide an equally good level of service as the inefficient fitting.

Not all water-efficient technologies accord with the 'fit and forget' philosophy. For example, rainwater harvesting and grey water systems both require the maintenance of tanks and filters. They also require the fitting of alternative piping mechanisms in the house – so the 'fitting' process is a substantial replumbing job – and the existence of this dual water supply needs to be understood by anyone carrying out subsequent maintenance on the house. While they remain 'unusual', such systems may impact on the resale potential of a house because they create anxieties about maintenance for potential purchasers. In this respect, such systems are unlikely to be fitted in a UK context by anyone who lacks access to plumbing expertise or who is not intending to live in their property for the long term. It is notable how these influences change if the fitting of such devices is becoming more of a norm in a society, such as in many cities in Australia (Sofoulis, 2015). Here, plumbers are increasingly knowledgeable about dual pipe systems. In these respects, we can see that what behavioural theorists may represent as an individual 'choice' is actually tied up in broader community understandings about plumbing.

At present, in the UK, while efficient technologies do exist it is only the intrepid consumer who seeks them out and ensures that they are fitted. Most suppliers of (for example) bathroom equipment market devices on their appearance and/or the anticipated pleasurable sensations arising from their use; meanwhile, information on water and energy efficiency is lacking, or hidden in the small print. Such approaches can be contrasted, for example, with the USA (Speight, 2015), where federal rules require universally higher levels of water efficiency, or with Australia, where the market for water-efficient devices is better developed.

Financial incentives

Financial incentives are ways in which pricing ensures that the users of a service contribute to its real (economic and maybe environmental) costs. Most developed countries have water meters in homes, and hence apply a 'metered charging' strategy in which customers pay a higher water bill when they use more water. Typically, one component of the bill relates to water supply and another part to waste water disposal. (Note, however, that one water supply meter usually measures both, as the quantity of water extracted is as assumed to be minimal.) There is usually also a fixed element of the bill. Paying for the drainage of surface water is a further element of cost that may, variously, be hidden within the other charges, listed as a separate fixed charge, or varied according to the extent of impermeable surface.

The 'typical' system of paying for water described above varies from an older system of paying for water in which a fixed charge for water services depended on the size of the property. The latter fixed charge system is still partly used in the UK, Canada and New Zealand. Some water utilities also have pricing systems that allow access to a small amount of water use for a low price, but then raise the cost for use exceeding a specified quantity. Such 'rising block tariffs' offer relative subsidies for low water users (Defra, 2008: 80), and are intended to enable access to water for those on low income. A further approach to water tariffs varies the cost at peak times – for example, during the summer.

The 'fair' or appropriate way to pay for water is a matter of debate (Bakker, 2004; Kayaga *et al.*, 2003; Page, 2005; Page and Bakker, 2005). It is therefore worth briefly exploring some of the different arguments made about this issue. These debates relate to: (1) the overall nature of costs of centralised water service provision, (2) the individual costs of centralised water provision and (3) the share of benefits of water service provision.

Water falls out of the sky and therefore, according to some perspectives, it should be free to everyone. However, the process of collecting water, treating it and distributing it to homes has significant costs relating to the construction and maintenance of infrastructure. Likewise – though less often discussed – the collection and treatment of waste water and the collection and disposal of surface or storm water requires similar expenditure. Typically, the ongoing cost of infrastructure provision is relatively low compared to the initial investment and/or the extra cost of providing for a little bit more service during a period of peak demand or scarcity. The high fixed cost of infrastructure provision stimulates debate about how it is best paid for.

One approach to infrastructure provision would be to expect all users or potential users to share the fixed costs of centralised infrastructure provision. This perspective might lead to state support for infrastructure provision, or equally for a fixed-price model of charging. Under either system, the water utility becomes incentivised to promote water saving, both in their part of the physical system and among water users, particularly if their water supply capacity is close to the limit where further water abstraction causes irreversible ecological damage. However, the disadvantage of such a system is often argued to be that the water users have no financial incentive to reduce their water use.

The contrary argument, that the variable price of water should be high for the user, is often employed when authorities are trying to encourage people to save water. This perspective, which focuses on the individual costs of centralised provision, suggests that people will put more value on and take more care about a resource for which they have to pay highly. However, the evidence indicating that people actually change their water use according to cost is variable (Jeffrey and Geary, 2006: 311). This is perhaps not surprising, as water is a relatively low part of most people's monthly income and (unlike food) they do not usually pay at or near the point of consumption. The only exception arises for those on very low incomes for whom water can be a significant cost.

Those promoting metered charging for water often suggest that the costs of supplying water should be borne by those who use it. This argument does not take account of how the cost of water service provision varies widely between properties and over time. For example, it is more expensive to supply water to homes at the top of hills, where the water must be pumped, compared to those at the bottom of hills, where water can be supplied by gravity. Working out how these costs vary between households would require new and different data than is commonly used today. Moreover, because of concerns about the inequity that would be associated with high costs in difficult-to-provide areas, such differentiation of water costs within a water company area is not generally considered politically acceptable. (In contrast, the variation in tariffs between water company areas is not seen as a problem.)

As well as geographical variation in tariffs, temporal matters also impact on the costs of supply. Hence, there is a logic that argues for diurnal or seasonal tariffs, with high prices coinciding with times when the system is close to capacity. In the past such variable pricing would not have been possible, but, as smarter meters and digital communications become more common, such variable pricing becomes technically viable. In relation to charging for electricity, Australian utilities have experimented with variable charging at peak demand periods. Interestingly, research on the mechanisms for change during these experiments suggest that just asking people to reduce demand during peak periods may be as or more effective than changing the price (Strengers, 2010).

Another way to consider the issue of how centralised water services should be paid for is to consider who benefits. As noted in the discussion of water practices in Chapter 4, water services confer both individual and communal benefits. During an era (1750–1900) when there was much concern with public hygiene and with people smelling, the potential communal benefits of universal water service provision appear to have been more important, and, in the UK at least, central and local government provision of these costs was justified. At present, the successful normalisation of very 'high' standards of cleanliness means we give less thought to these communal benefits and hence see water as largely associated with individual gains. This emphasis on the individual benefits may be one explanation for the shift towards metered charging in the UK and elsewhere.

Compared to most countries, the UK is unusual in that metered charging is not the main way of paying for domestic water use. In 2016, 40 per cent of English householders were metered (UK Water, 2016), while the other 60 per cent pay a fixed charge related to size of their home. The number of homes subject to metered charging is slowly increasing, but it is expected to be 2050 before 90 per cent of homes are metered (OFWAT, 2011: 2). Most English water companies operate a single simple volumetric tariff for their metered customers. There have been periodic intentions to impose universal metering on household customers in the UK (Bakker, 2004; OFWAT, 2011), but these have floundered because of political concerns over the impact on high water-using customers with limited incomes.

In many respects, the mixed system of the UK can be seen as the worst of all charging systems. This is because clear communication about water saving is hindered through the differential tariff structures experienced by UK households. If all householders were metered, then clear messages about 'we can all save water and money together' would be appropriate. If all faced fixed water charges, then messages could concentrate more on the potential harm to local and national environmental assets.

This section has focused on how financial incentives are seen as a way to encourage direct water saving, highlighting the range of potential tariff structures for water supply and the choices about what is seen as 'fair'. As has been demonstrated, issues about how water saving is incentivised also relate to broader debates about how centralised water infrastructure is valued and paid for.

Persuasion

Providing information to persuade customers to use less water is a complex strategy. Information can take a number of forms, depending on the information suppliers' intentions: it may be motivational – for example, highlighting the local or national harm that water use is doing; it may be instrumental – for example, informing the customer about actions they can take to reduce their water use; or it may be personally informative – for example, providing individuals with information about their household water use through time. To be effective, any such strategy must first access the public; it must also motivate them to save water, as well as providing them with the knowledge and ability to use less water (Sharp, 2006). Typically, persuasion is used alongside other strategies. For example, information on saving water in the home may be delivered to customers alongside a cistern displacement device (an object to place in the toilet cistern to reduce the quantity of water used), or at the same time as a switch to metered charging (see Knamiller and Sharp, 2009, also the discussion of demand management in Thames Water from Chapter 3).

A critical factor that impacts on when and whether people pay attention to information provided is their level of trust in the body providing the information (Knamiller and Sharp, 2009). Such trust is not just generic; it also depends what information is provided, whether the source is seen as authoritative and unbiased on this topic and on whether the communication itself is seen as authentic. In the UK, for example, water companies are viewed with suspicion in relation to anything that might impact on their ability to make money.

Information seeking to support changes in water practices tends to be more successful if asking people to participate in an activity that is already being carried out by others. For example, Sofoulis (2011) discusses an information campaign by Brisbane Water that sought to encourage everyone in the city to reduce their water use per person to a norm of 140 litres per day. Unlike water restrictions, this approach treated the water user as able to make informed choices about their own water use and in what ways it could be most easily reduced.

Critiques of behavioural approaches

The understandings and measures described above are all – implicitly or explicitly – informed by a behaviourist perspective. Shove (2010) has characterised these behavioural methods as 'ABC' approaches, in which influencers seek to understand 'A', Attitudes, and B, Behaviours, in order to achieve C, Change. Some key characteristics of these approaches are that:

1. Their central focus is on the total water use demanded from the supply system. In this context 'demand management' becomes sloppily broad: it includes the leakage demands of the pipe system (Sharp *et al.*, 2011b), while the total amount supplied per household can be mistaken for average household use (Sofoulis, 2011: 800).

2. Bill-paying customers are seen as independent decision-makers who choose to use water (or not) in various ways, and who are implicitly able to exert complete control over their household. The government and water utilities are seen as external to the consumers; they are enablers who can help these independent decision-makers choose the 'sustainable' path (Shove, 2010; Sofoulis, 2011; Strengers and Maller, 2012) through rules, technology, information and incentives. Developments in culture or the context for decision-making are also regarded as external to the individual decision-maker (rather than individuals being recognised as part of a community) and hence can be ignored as a complicating factor.
3. Demand management is perceived to require generalisations about the population, their behaviours and their attitudes. As depicted above, in this context there is great focus on the average water user (actually the average household) and their per capita consumption, both in total and as split between micro-components. There is very little consideration of the variation in water use between and within households. Insofar as consumers are differentiated at all, it is the bill-payers who are split into market-research driven segments, usually using socio-economic criteria that are largely unconnected with water use. (The work discussed above by Icaro for Defra forms an example of this approach.)
4. The approach promotes the testing of different techniques for achieving changes to behaviour in order that an evidence base can be established before they are implemented more widely. This assumes that it is appropriate to conduct social experiments on the public, and also that techniques can be unproblematically down-scaled (in order to conduct an experiment) and then up-scaled (if the experiment proves successful) (Shove, 2010; Sharp *et al.*, 2011b; Fam *et al.*, 2015).
5. While attempts to address demand have, at various times, focused on altering the activities of households, water suppliers and water fittings manufacturers, the combination of these practices is seldom viewed as a connected system that creates the conditions to produce and reproduce current levels of water demand.
6. Insofar as measures focus on households, relatively little attention is paid to apparently 'small' matters such as who contacts them, when and why, how are they asked to participate and what are they asked to do. In this respect the crucial micro-politics of water efficiency initiatives are given little attention (and yet may be key to their relative success) (Sharp *et al.*, 2015).

It is in the context of these critiques of behavioural approaches that practices perspectives on water management have emerged.

Understanding water use practices

During the first decade of the twenty-first century a new set of 'interpretive', 'cultural' or 'practice' studies have sought to develop new ways of thinking about water use (and also energy use, food use and waste production). Examples of such approaches pertaining to water use include Shove 2003; Southerton *et al.* 2004; Strang 2004; Knamiller and Sharp, 2009; Chappells and Medd, 2008; Bell and Aiken, 2008; Sofoulis, 2011; Doron *et al.*, 2011; Strengers and Maller, 2012; Browne *et al.*, 2013a; Pullinger *et al.*, 2013 and Fam and Lopes, 2015. Taking as their starting point the observation that people do not make explicit choices about water use behaviours, the research has used observation, questionnaires and interviews, as well as diaries and focus groups, to examine how and why people use water in households, and hence to develop a new conceptualisation to inform the way governments and water utilities interact with the public. Household water use practices are understood as local, variable, differentiated and linked closely to previous experience/culture and available infrastructure (Fam *et al.*, 2015).

An example of an interpretive approach to understanding water practices is the work done by Sofoulis and colleagues in a Sydney suburb during a period of drought and water restrictions as consultancy for a housing developer (Sofoulis *et al.*, 2005). Using a combination of interviews, water diaries and a questionnaire, this investigation studied water cultures among the residents of western Sydney, with a particular focus on whether and how people recycled water. The study showed that respondents were very aware of the value of water and that many had made sacrifices in terms of changes in their indoor and outdoor water use practices in order to save water during a drought period. About half of the participants had undertaken some informal grey water recycling activities, mostly with buckets and usually oriented towards maintaining the health of valued gardens. Three types of water saver were tentatively identified: older people willing to conserve for themselves at the cost of time and effort; younger people willing to install technology enabling them to save water; and sustainability aspirants who want water and energy efficiency fully integrated into their homes. The recommendations focused on how to meet people's perceived need for help understanding, accessing and fitting water-saving technologies – in particular, to make use of their 'wasted' grey water.

A second example is the work done by Woelfle-Erskine (2015) in the Salmon Creek catchment on the Californian coast. A small water company (supplying 39 properties), along with numerous private wells and surface water withdrawals from the creek, had reduced its flow to critical levels such that, in 2009, the Salmon for which it is known were unable to spawn. In response to this, a programme for installing subsidised water tanks was devised to reduce water withdrawals and maintain groundwater levels. In 22 research interviews, householders took the researcher to their water source (whether tap, tank or well), with discussion focusing on water knowledge and ideas about responsibility. The research found that the programme had developed and enhanced the detailed knowledge that many residents already had about the water in their neighbourhood. There was a snowball effect in that people learnt about rainwater harvesting from the early adopters, and then, schooled by their friends, applied for a rainwater tank themselves. The research identified that while there had been a recent shift such that collective management of the catchment's water could be discussed and explored, implementation of collective change beyond the individual and voluntary (e.g. strategic development of infiltration basins, further rules and enforcement about water withdrawals) had not yet been enacted. Woelfle-Erskine concludes that residents managing their own water system became more interested and more entangled in water management and its consequences for their human and non-human neighbours.

As well as yielding rich findings to inform local water managers, these case studies also drew conclusions with wider applicability: for example, Woelfle-Erskine's observation about the effect of managing your own water. Indeed, these cases are excellent examples of the sort of 'narratives' discussed in Chapter 1, which promote thought and yield understanding of diverse water situations. The cases, additionally, illustrate some of the themes and issues that can inform research questions and processes that might help in the design of water-related qualitative research. Nevertheless, such case studies do not always count as evidence in contemporary water utilities (Fam *et al.*, 2015): they are seen as small and particular, and hence not offering general lessons. While I reject this critique, it does pose a broader question about how these interpretive understandings can best support action in the 'scaled-up' contexts that utilities are concerned with. Two examples where researchers have scaled up interpretive research are given below.

The meta-analysis conducted by Lesley Head and her colleagues on cultural responses to climate change in Australia provides one example of how interpretative case study research can be seen as more broadly applicable (Head *et al.*, 2016). Drawing on the methodologies used for meta-analysis in other disciplines (e.g. medicine), a series of relevant practice-related case studies have been systematically identified and analysed for general thematic findings. They conclude that adaptiveness of households is pervasive, and tends to build on previous experiences – for example, water or energy saving is informed by people's experiences of rural childhoods. They suggest that part of the usefulness of their work is to triangulate the findings of the different studies against each other in a way that is both more systematic and more geographically focused than in general discursive reviews of the literature. As such, they suggest their study might provide a strong basis for an international comparison. Head and her colleagues also argue that their work can support the design of interventions: household capacity for adaptiveness is clearly a resource to be drawn on and utilised, but, likewise, themes about autonomy, privacy and concerns about government control provide 'warning bells' against which interventions in Australian households might be checked.

The work of Alison Browne and colleagues has sought to draw on interpretive techniques, but in a wider context offering more room for generalisation, and hence offers another approach to overcome the criticism that interpretive and practice approaches cannot be generalised (Browne *et al.*, 2013a; Pullinger *et al.*, 2013; Browne *et al.*, 2013b). In an investigation that involved both detailed questionnaires and interviews, this research sought to understand how people from the South East of England undertook water-using practices – specifically, the practices of washing themselves, gardening and laundry. Some aspects of the investigation drew on standard quantitative research practices – for example, in the selection of a sample of the population from census areas, encompassing a representative balance of different levels of deprivation (Pullinger *et al.*, 2013: 13), and the use of a detailed questionnaire, albeit one which was conducted in an interview format. However, in contrast to the attitudinal work discussed above, a crucial technique was cluster analysis, which was used to identify 'groups of practices' undertaken by different sets of people, highlighting different types of practice in relation to three areas.

As an example of where this analysis leads, Table 5.1 summarises the results of this work in relation to personal washing, listing six different practice clusters that were observable within the population. The final column of Table 5.1 shows how the cluster analysis leads to particular questions and potential routes for intervention for administrators concerned with total water use. For example, given that 40 per cent of the population are 'simple daily showerers', it is clear that any intervention that successfully reduced the amount of water used in a typical shower would have a significant impact on total water demanded.

Compared to behavioural studies, practice approaches are much more concerned with how, when and by whom water is used, an approach that is strongly supported by cluster analysis such as that carried out by Browne and colleagues. A key potential advantage of such perspectives is that they can support and guide intervention strategies, which may be focused on consumers but may equally be focused on those designing or running infrastructure.

Some examples of ways in which cluster analyses of practices (such as that carried out by Browne and colleagues) could help target intervention strategies include:

1. Focusing on a specific problematic practice – for example, more on daily showering or heavy garden watering rather than the overall consumption level.
2. Linked to 1, targeting the specific segment of the population who commonly undertake a problematic practice – for example, young people who may be more likely to shower more than once a day, or people of all ages who invest time and energy in their gardens.
3. Likewise, targeting marketing media, locations and metaphors according to the segment of the population of interest.
4. Focusing on the public at particular key life stages – for example, when young people move out of the home, or when people move house – to inform and support individuals at these key 'change' moments when new practices are developed and embedded.
5. Using infrastructure suppliers (such as bathroom manufacturers, DIY stores or garden centres) to help develop infrastructure and marketing concerned to support ways to tackle problematic practices.

From a practice perspective the behavioural studies that sought to link total water use with physical or demographic characteristics of a household are inappropriate because total water use is a very poor proxy for the combination of different practices that are executed. Specific practices can be seen as the missing causal link in this chain, with high household water use resulting from a combination of high

TABLE 5.1 Different types of washing practices

Clusters	*Description*	*% pop*	*Socio-demographic characteristics*	*Questions/potential interventions*
Simple daily showering	Very standard washing pattern for convenience	39	More likely employed and affluent and healthy. More likely to have water meter	Help power showers be more efficient. Successfully encourage shorter showers
Out and about washing	At least daily, in and out of home. Leg shave for women. Motivated to help environment	16	Younger than average, full-time work, more male. Healthy and likely to live in house of three or more occupants. More likely to have water meter	Similar to daily showering. Is this a 'young and active' age effect, or is there a concern that this group's higher water usage will be maintained as they age?
Attentive cleaning	Shower or bath at least daily with high emphasis on personal grooming, including shaving	15	More likely to be young, female and to have children. Likely to be healthy and not retired. Less likely to have water meter	Relatively higher-frequency bathing and use of baths suggests possible intervention – needs to plug into trends as this is a fashion-conscious group
Low-frequency showering	Less frequent washing usually or always showers, mostly at home	12	Older and more likely to be retired, living alone or with one other. More likely to have health problems. Less likely to want to do more to help the environment	Relatively low level of water use probably. But encouraging others to adopt these practices may be challenging
High-frequency bathing	Always having baths at least six times a week, usually motivated by cleanliness	11	More likely to be female, unemployed, rented accommodation and less likely to have a water meter. Sometimes without shower in house	Potential to target fitting showers in rented accommodation
Low-frequency bathing	Always bath at maximum five times a week, motivated by cleanliness, with baths low	7	Older than average, consequently less likely to be in full-time work, more likely to be retired. More likely to be renting and less affluent	Probably a 'relic' from previous bathing practices. Relatively low water use – though, if we could find a way to channel these behaviours to others, that would be interesting

Source: Browne *et al*., 2013b

water-using practices. By looking in detail at what practices are undertaken, and by considering these alongside household physical, demographic and cultural characteristics, Browne and her colleagues have begun to gain more understanding of who uses large quantities when and why.

As the above discussions have shown, the practices approach promises to offer some new insights as to how to understand the processes of water use and hence to potentially support and inform new sets of interventions designed to achieve changes. As the points above make clear, the specific interventions that the practices approach might advocate may coincide with those previously used by behavioural work. The key difference between the practices and behavioural work relates to three things: (1) understanding of the context in which the practice occurs (meanings, material and procedures); (2) targeting of specific practices and those linked to them; (3) the development of a combined programme of measures which seeks to work in the context of the whole water system, rather than focusing exclusively on the water user as the agent of change.

What would water conservation look like if informed by practice theory?

Practice theorists acknowledge that while they have a great deal to say about how practices should be understood, the precise implications of practice thinking for those who are promoting sustainable lifestyles are not always clear (Spurling *et al.*, 2013). In this section, I try to go even further to examine the question: what are the implications of the practice perspective for a water utility seeking to develop a more sustainable water system? What follows combines my knowledge of the water sector, my observations of behaviour change campaigns (Sharp *et al.*, 2015) and my reading and understanding about the practices perspective (drawing on Browne *et al.*, 2013b and Spurling *et al.*, 2013, among others). It can be seen as a set of steps for utilities intending to mobilise communities to change practices.

1. Identify motivations

A water utility wanting to take action on water needs to begin with clarity about their motivations for seeking change. In the past, water conservation campaigns have tended to focus on the generic goal of reducing aggregate water use, but this is often a poor proxy for dealing with more specific issues. Water conservation may address a number of water-related issues including; seasonal/diurnal pressures in the local water system; regional water stress; locations with high company costs (for example, where a large amount of energy is used to achieve water supply); enhancement of communications and exploration of service-level expectations with their customers; the undertaking and demonstration of corporate social responsibility in relation to climate change and fuel/water poverty. In this way there are likely to be a number of different linked problems that might be addressed together through working with the public: some are acute and geographically/temporarily localised; others are more generic and run across the water utility area.

It should be noted that local differences in a utilities' network may mean that they face real cost advantages to targeting water savings campaigns to specific localities/time periods. They should not be shy of following such advantages; unless they have geographically differentiated prices, savings in one location are likely to achieve financial benefits for all. Also, targeting action in a specific area where there are large savings to be made can act as a pilot for subsequent scaled-up activities intended to cover a larger area. Finally, there is an argumentative strength if a utility has a good understanding of why they are undertaking specific actions; asking consumers to look in detail at their motivations and practices is more easily done if the water company can show that it has done the same.

2. Investigate practices

Utilities should then explore what practices contribute to the identified problem. These might include their commercial practices such as leakage from pipes, as well as customer practices, such as a high demand for water at a specific time/place (e.g. on a daily basis between 6am and 8am). Investigations of consumer practices should begin by asking what is done at the critical time, before examining the meaning of the practice(s) to local people, and where, when and how they do it (Spurling *et al.*, 2013). The research might additionally ask about the context for the practice: is it part of a sequence of other practices, and, if so, what are they?

Community partners can be a very important means of exploring local practices because of their access to communities and understanding of local cultures. Examples of such organisations include community/residents associations, shared interest-based activity groups (e.g. Scouts or mothers and toddlers), faith-based groups and schools. By working with these groups (and perhaps providing an incentive contribution to their organisations), communities can be invited to contribute to focus groups or small-group interviews, aimed at better understanding a practice and the potential to achieve changes in it within a specific local area.

The outcome of understanding a practice will be some ideas about different ways in which it might be changed. This might involve seeking to recraft the practice, substituting alternative practices, or changing how practices interlock (see Table 5.2) (Spurling *et al.*, 2013). Thinking about influencing a practice will also involve thinking about the different organisations and social norms that contribute to the way it is currently carried out – organisations that themselves might be considered as partners allied in action to change the practice. The lesson from Woelfle-Erskine's (2015) study of water tanks in California is also relevant here; getting people involved in understanding a problem is also part of the process of engaging them in how it is addressed.

3. Assemble an alliance and form a programme of action

Campaigns to influence practices are most effective if they are done in an alliance. By working in partnerships, the utility cannot only access others' expertise, but

TABLE 5.2 Ways of changing a problematic practice using the example of high water demand from a morning showering

	General explanation	*Example*	*Potential partners*
Recrafting a practice	Seeking to change the meanings, materials or procedures associated with a practice	Supporting people in achieving lower water use from showering through fitting low-flow shower heads	Equipment manufacturers, DIY stores, toiletry manufacturers
Substituting a practice	Identifying a more sustainable practice that can achieve the same benefits for the user but is more sustainable	Promoting the use of a daily flannel wash rather than showering as a means of morning ablutions	Toiletry manufacturers
Changing how practices interlock	Changing infrastructures and institutions can shift their sequence and interrelations to be more sustainable	Reducing 'grooming expectations' for workplaces could free many from daily showering expectations, as well as enabling other changes such as active daily commuting	Employers and city transport planners, public health authorities

Source: Developed from Spurling *et al.*, 2013

also the credibility of their external organisations can add to the effectiveness and draw of any campaign. Depending on the precise issue to be addressed, water utilities should consider working with local environmental groups, energy suppliers, energy savings organisations, or organisations concerned with resource poverty.

To develop a programme of actions, ideas about how to change practices that arise from the research need to be combined with understandings of different organisations' motivations, drivers and timespans. To bring an alliance of organisations together, however, shorter-term quick-win actions should be developed alongside one or two actions that will require more work but that can be pursued through partnership over a series of years. The programme of actions is likely to include a number of actions already underway (for example, utility action to address leakage) and longer-term actions that might achieve a permanent change in practices, as well as immediate activities to achieve involvement, publicity and a feel-good factor.

Insofar as activities involve household-level interventions, these might be expected to be more effective if:

1. Existing networks of trust are utilised and expanded in seeking to find means to involve people and alternative means to take action (for example, the utility

might work with the community organisations through whom they did the initial research).

2. Those asked to take action should understand their contribution as part of a wider set of collective actions and should be recognised for contributing to the common good.
3. Monitoring as well as ongoing discussion with all parties should reveal whether aspired-for changes have been achieved, as well as any unexpected positive or negative effects, and this information should support the process of refining and developing the initiative.
4. The role, contributions and changes made by those organising and encouraging the action are emphasised in the publicity as well as those that are aspired for by the public – the public should not be made to feel that they are acting alone.
5. Processes of learning and reflection should enable those organising the initiative to identify and learn lessons so that it can be improved.
6. This process of water utilities helping their customers to learn about and potentially interact differently with the water system may be best combined with the reverse process – in which the utility learns about and potentially acts on its customers' priorities and concerns. In this way, mobilisation may be combined with participation.

In the longer term, programmes of actions such as those discussed above are more likely to be possible if funding processes recognise the benefits of programmes achieving multiple 'wins' across different areas of policy.

4. Measuring the success of mobilisation

Connected practice-based attempts to influence water use cultures complement technical changes in the armoury of methods that can be used to address contemporary water challenges. However, crucially, the 'success' of mobilisation should not be judged through quite the same yardsticks as are customarily used to assess technological investments in water. These are typically focused on short-term measurable changes; for example, the quantity of water 'gained' through investing in new pipes in one district is contrasted with the amount lost in an equivalent area with no investment. Care needs to be taken when applying such assessment processes to investments in mobilising public action because:

- Water organisations are relatively new to mobilising public action and are still learning how to do it.
- Investment in mobilising public action cannot be controlled in the same way as technology. For example, a hard-hitting regional campaign for water saving during a period of drought will be received by all living in the region, and it is not possible to cut off a specific neighbourhood as a control case study.
- People may object to feeling like they are part of a social experiment and, hence, react strategically if they discover they are being studied, such that the experiment is invalidated (Knamiller and Sharp, 2009).

- Social equity and ethical concerns may mean that the sponsor organisation is unable to identify a control region.
- Investments mobilising public action are at least partly cumulative: messages may need to be heard a number of times, investments in household infrastructure may have to await a period of redecoration, or a generational turn-over.
- Evaluation needs to recognise a range of possible benefits, not all quantitative, and some of which may not be known at the start. Hence, measuring the success of rainwater tank programmes through the storage capacity added to the system is to ignore the widespread learning and engagement that they seem to engender (Sofoulis, 2015; Woelfle-Erskine, 2015).

One means of evaluating mobilisation initiatives might be to draw inspiration from approaches used for considering participation. In a review of techniques for evaluating participation in resource management, Carr and her colleagues (2012) highlight widespread use of very short-term process measures (e.g. number of people participating) and very long-term outcome measures (e.g. quantity of resources preserved), but rather limited use of intermediate outcome measures (e.g. trust built or institutions changed), which they consider as providing a useful bridge between the immediate and the long term. To date, mobilisation techniques have frequently been measured alongside technical innovations according to their delivery of long-term outcomes measures. While, of course, these long-term goals are important, they offer limited opportunities for learning, particularly if mobilisation is measured in terms of its immediate effects on these long-term goals. The potential for collective learning about mobilisation initiatives is more likely to be realised if we also employ short-term process measures and intermediate-term process and outcome measures.

Contemporary action on water conservation in the UK

The above descriptions can be contrasted with the contemporary state of attempts to manage water demand. The UK is discussed not as a 'good' or 'bad' example, but rather as a means to demonstrate the complexity of water demand governance.

In the UK, it is government policy to promote a more efficient use of water (Defra, 2008). The policy document *Future Water* (ibid.) identifies adaptation to climate change as one of the key factors driving this concern. In practice, EU directives concerning habitats and water quality require reductions in the extent of abstraction. At the same time, the population is growing, particularly in the more water-scarce south-east parts of the country. As the government's agency with responsibility for the natural environment, the Environment Agency is an enthusiastic promoter of this priority – for example, through their quarterly *Water Demand Management Bulletin*.

The water companies' economic regulator, OFWAT, has somewhat more mixed interests in relation to water efficiency. Arising from the era of privatisation, OFWAT's primary interest has been to promote competition in the water industry, and hence to represent the needs and views of the consumer. While this interest has been modified and developed over the years, it can be argued to still be an overriding emphasis of their policy. A short-term approach to consumer needs would emphasise water prices, and would hence put little value on the promotion of water efficiency. A more long-term perspective recognises the value of water for habitats, the environment and also for industry and well being, and hence favours some promotion of water efficiency, particularly insofar as there are diurnal or seasonal pressures on water supply infrastructure. During the past ten years OFWAT has moved to press for water efficiency measures more explicitly. During the early 2000s, the requirement on water companies was purely that water efficiency was promoted, with no specific goals about achievement. Recently this has changed and, since 2010, water company 'water efficiency' targets have involved two main components (OFWAT, undated). All companies must achieve a base level of efficiency saving – that is, an average saving of 1 litre per property per year. They are expected to achieve this saving as part of their core business, as well as to contribute to the development of water efficiency knowledge – for example, through their contributions to funding the not-for-profit body Waterwise. Additionally, water companies are encouraged to undertake specific water efficiency schemes that they can negotiate with the regulator in terms of the potential investment needs and gains. The long-term intention is to reduce the average per capita use in the UK to 130 litres per person per day, which may be a challenge, given a current average of about 150 l/c/d and a history of this increasing by 1 per cent per year (Waterwise, 2013b; CCW, 2015a). Alongside efficiency goals, water companies are now expected to negotiate the outcomes they expect to deliver through interactions with 'customer challenge groups'.

Table 5.3 shows the different incentives exerting pressures on water companies in terms of promoting more efficient water use by their customers. It demonstrates something of a mixed incentive system.

As Table 5.3 demonstrates, water companies could benefit from being able to rely on consumers reducing their water use in response to drought. Additionally, in particular locations and at particular times, the water supply network is working at capacity. Outside of such 'peaks' the water companies' interests are more ambivalent. Insofar as metered charging exists (approximately 40 per cent of UK households at present (UK Water, 2016)), there is an apparent disincentive for the companies to promote reduced water use from metered customers, for this would also reduce their income. In the longer run, however, companies are compensated for any loss in income resulting from water efficiency gains through the 'revenue correction mechanism'. Moreover, in relation to those customers who pay a fixed charge for their water, successful promotion of water efficiency to those who are not metered would achieve savings for the companies. In addition to these 'income effects', it is important to understand that water companies are valued in the capital markets according to assessment of their assets. If there is sufficient pressure on their assets, then the regulator will permit rises in prices to support development of new 'assets' (that is, new reservoirs or boreholes). This adds to the companies' value in capital markets. In this respect the companies have an incentive to increase water usage.

TABLE 5.3 Factors influencing English water companies' financial interests in terms of water efficiency promotion

Factors encouraging companies to promote customer water efficiency
Could save money if customers respond to requests to reduce demand during drought periods
Could save money if customers respond to requests to reduce demand during local diurnal peaks
Successful water efficiency promotion would save costs while maintaining income from fixed-charge customers
Regulator promises to allow increases in prices to compensate for losses of income if metered customers become more water efficient through the 'revenue correction mechanism'
The regulator sets water efficiency targets (in terms of per capita use by domestic customers), and companies suffer a charge if they do not comply
Regulator insists on seeing action to promote water efficiency before investments in new assets are allowed
Factors discouraging companies from promoting customer water efficiency
Successful water efficiency promotion reduces income from metered customers
Capital value of company increases if demand is sufficient for regulator to permit investment in new assets

Hence, the water companies' interests are more complex than those of the government as they relate to usage during peaks, to pricing structures and to capital values. In practice, the water companies' approach appears to be focused on complying with the regulatory expectations about promoting water saving, but highlighting the potential need to develop new assets where stocks are low.

Other organisations that sometimes have an interest in water savings are local authorities. In the dryer south and eastern parts of the UK there is a fear that water resource shortage could threaten the need to expand housing and jobs in the areas. Some local authorities (for example, Kent County Council) have taken an active part in exploring and promoting water saving in their area (Sharp *et al.*, 2015).

On the whole, the British public does not prioritise water saving. Not only is water a small part of the household budget, it is usually paid by direct debit and involves little engagement with the quantity of water used. Moreover, the general British perception is that we live in a wet country and have no need to save water. This perception contrasts with the view of the environmental regulator – and the requirements of EU legislation – that stress the need to reduce the extraction of water from water bodies, particularly in the south and east of England. Furthermore, as the practices research stresses, water use is a habitual part of our daily lives and, hence, water practices are seldom perceived by the public as a matter of choice.

In a fascinating account of the changing approach to governing water demand in the UK, Walker (2014) describes the chronology of changing initiatives and incentives in relation to water use. For Walker, it was the narrowing definition of scarcity – as based on a catchment, rather than seen in aggregate – which stimulated the crucial change in water organisations' remit in the late 1980s. From the privatisation of the UK water companies onward, then, the institutions of water management began to be seen as partly about providing a good technical service but also about ensuring demand management. Successive governments have spun water company regulation in a slightly new way, but each works within the legacy systems allowing limited change. Hence, an emphasis on light-touch regulation and the market allocating resources under the Conservative governments of 1989–97 was followed by a period of regulation – for example, leakage and water efficiency targets – under Labour (1997–2010). More recently the coalition and Conservative governments have sought to create the conditions in which the market can operate more fully – for example, by removing water companies' monopolies in the supply of water to commercial customers.

Walker's account is particularly useful in highlighting the moral dimensions and questions arising from seeking to influence water use. As he stresses, water utilities have evolved in a context in which particular practices are deemed as legitimate or normal. In attempting to influence these evolving practices, their actions have been portrayed 'as a threat to public health, a compromise in the modernisation of living standards and economic growth, and more recently as an infringement on consumer sovereignty' (ibid.: 407). Any further attempts to influence water practices – whether by utilities or others – is likely to encounter similar criticisms.

Walker argues that, even by the industries' own measures (related to leakage rates, per capita consumption and domestic meter penetration), water conservation measures in the UK have failed. Most of all, he emphasises how 'as the remit of the sector has broadened, so too has the demand for an institutional structure capable of governing a complex social coordination', requiring 'a level of cooperation between water companies, consumers, and intermediary bodies such as property developers and appliance producers which the industry has struggled to support' (ibid.: 405).

The discussion from this section supports Walker's conclusion. Overall, there is little evidence in the UK of any collaborative sense of urgency in developing a more water-aware approach. A number of different organisations – the water companies, Waterwise and OFWAT – are taking *some* actions, but these do not add up to a coherent programme of activities that are connected and logical for the public. Communication and action is further hampered by the mixed legacy of charging systems. This lack of urgency or concern contrasts with the Environment Agency's designations in relation to water scarcity and over-abstraction in the dry south and east of the country.

While this section has focused on the promotion of water efficiency in a UK context, the relative newness of water efficiency promotion in the UK is paralleled in many other locations and, indeed, by other similarly 'new' watery topics of governance (e.g. flood resilience or water quality). Hence, many of the conclusions – for example, regarding the limited and uncoordinated promotion of water efficiency – are likely to apply in other spheres of water activity and in other geographical locales as well.

Conclusion

The context for this chapter is growing concern with water supply scarcity and the need for water organisations to take on a broader remit that works with and influences other organisations and individuals to achieve a cultural change in quantities of water use and expectations thereof. Armed with knowledge from Chapter 4, we can begin to be assured that people's water use has changed in the past and is likely to change again.

Growing out of water forecasting techniques, initial scholarship and action to consider whether and how the public could be influenced in relation to their water use were based on behavioural models. These sought to relate people's water use to measurable factors such as their social class, gender and age, as well as to the water-using features in their homes. Developments of the behavioural approaches took on the importance of exploring people's attitudes to water use. Even though recent explorations of attitudes often embraced qualitative methods, all of the behavioural approaches were essentially positivist in their perspective on humans, who were viewed as isolated individuals taking independent decisions about their water behaviours.

Scholarship on 'practice theory' has developed a critique of behavioural approaches, highlighting the importance of how people are conceptualised. If current explorations of water practices are further developed and applied, they promise an approach to understanding and influencing water use that is both more complex and holistic. Instead of seeking to change the per capita consumption of an imagined, average, utility-maximising customer, rather it is specific problematic practices that become the focus of attention. In recognition of the variety of factors that influence practices, campaigns are likely to include the organisations and places that facilitate the practice (e.g. power shower manufacturers), as well as those who carry it out (e.g. people who have power showers). The approach would be likely to recognise that practices are formed by and in a community, and that it is the culture of the whole community that needs to be influenced. The campaigns would also be likely to 'join up' a number of agendas (and agencies) that relate to the practice – for example, 'power showering' might address water scarcity and energy efficiency.

As the account of water conservation in the UK has shown, practice perspectives introduce an institutional challenge. There are complex tangles of existing organisations and institutional relationships which all influence the practices that use water. In seeking to embrace a holistic understanding of different linked problems, a set of central questions is raised concerning which factors we want to see as integrated, and how we support and enable flexibility rather than setting up another fixed system of relationships. Moreover, insofar as we do move forward to a more joined up and sophisticated approach to managing water problems, questions are raised about legitimacy and appropriateness of governance organisations seeking to mobilise action and to influence water cultures.

In terms of the progress this chapter has charted towards a more hydrosocial water management, the report card is mixed. On the one hand, the complex understanding of individual practices appears to provide a more nuanced and empathic approach to understanding and influencing how water is used than the preceding behavioural approaches. Exploring further what a practices approach could bring may offer a promising route towards more people-sensitive water management. On the other hand, the account highlights the challenge of designing institutions for managing water supply in the light of multiple overlapping water and non-water issues. The governance of hydrosocial water management is confirmed as a highly complex issue.

6

WATER QUALITIES

In the preceding chapters on water supply and water use in the home the discussion has focused on the 'clean' side of water management, asking how fresh water is delivered and used. In this chapter the focus turns to the complex issue of different waters' qualities.

The qualities of different waters can be seen as intimately linked to urbanisation and technocracy. Increasingly dense populations raise the challenge of handling more sewage and other waste water. Urbanisation also changes surfaces: vegetated rural land holds rainfall on plants and in the soil. In contrast, the compacted mud or paved surfaces of urban settlements cause rapid surface water flows that pick up and carry pollution. As we have seen, historically social norms carefully protected drinking water from the encroachment of pollution – for example, from washing tubs. But in the context of the overwhelming rapid urbanisation faced by the UK in the nineteenth century (and in many urbanising cities in emerging economies today) environmental waters – and hence, potentially drinking waters – become contaminated. We have already seen one response to this – the collation of drinking water from further afield. In this chapter I chart further technocratic responses, involving the separation of one flow from another and their management through large centralised infrastructures. But I also begin to explore ways in which this pattern is being or could be broken.

Water qualities are complex for a number of reasons. First, we attach strong cultural meanings to water. Water perceived as pure or natural is highly prized, while that perceived as dirty or polluted is repudiated. Water qualities matter to us, and matter highly at a deep, visceral, personal level. Second, in discussing water qualities, we necessarily tangle up the traditionally siloed parts of water management – the clean and the dirty. Our water management has evolved to treat clean water as separate from drainage and foul sewage precisely because not separating these different waters caused problems. But our management processes evolved in a particular time and place, and then (to a significant extent) have been universally applied across cities ever since. In exploring water qualities we have the opportunity to revisit these management practices. Third, there is a question about what *makes* good quality water. In Linton's account described in Chapter 2, the shift from pre-modern to technocratic management was associated with a change from 'waters' to 'water'. With this change, understandings about the many different qualities of waters from different sources were erased in favour of universal expectations about drinking water quality. This is matched by increasingly universal expectations about river water quality. Of course, these changes have been associated with significant public health and environmental benefits. Nevertheless, in reconnecting people with water, should we seek to move beyond these universal expectations? Already, when discussing the 'water sensitive city', the use of a variety of 'fit for purpose' waters is advocated (Wong and Brown, 2008). If we embrace the 'fit for purpose' concept, what variety of different waters would we seek to engage with? Would the benefits of universal 'clean' water be lost again? And, finally, as a layperson it is difficult to assess waters' qualities. While we can see or smell some aspects of (poor) water quality, many other aspects of quality are not discernible through our immediate senses, requiring expert input or testing to be evaluated. As the following accounts will show, this creates distinct power differences in which those with positions of authority measure quality, without necessarily exploring which aspects of water quality matter to those engaging with water in a particular circumstance. In this respect, monitoring technologies have given governance organisations a degree of power over individuals and households. In the future there is potential for low-cost sensors and other devices to make water quality more knowable: could this also make water management more democratic?

In exploring water qualities, then, the fundamental social questions probed by this chapter concern, first, what makes water quality – in other words, what is it that we measure and value to designate something a 'good' water? And, second, though closely connected, is the question of how we manage waters to ensure the qualities are appropriate. Finally, insofar as there are different potential ways of managing quality, what do the historic stories about learning to manage quality in a technocratic world tell us about a putative further transition to a more 'reconnected' water management?

These questions are explored through five encounters with water quality. The first encounter concerns the contemporary water supply and disposal system on the New Zealand island of Waiheke. The location is unusual because it is effectively a commuter suburb of Auckland and yet has neither a reticulated water supply nor any centralised sewer system. Instead, each property has its own system. The anecdote raises questions about perceptions of quality and the balance between communal and individual responsibility for public health. Following Waiheke, the discussion embarks on the first of three related accounts of the development of the London sewer system. In examining the shift from cesspits to the idea of a shared sewer system, the discovery of the waterborne nature of cholera and the construction of the great 'interceptor' sewers, the chapter recounts the dawning development of our contemporary understandings and physical systems for maintaining water quality. The stories provide three different 'slices' in the transition from individual to technocratically organised shared sewerage systems: the first concerned with ideas, the second with knowledge and the third with governance. The final part of the chapter examines attempts to address contemporary water quality in a 'typical' degraded urban catchment in the city of Bradford. The discussion demonstrates the ongoing challenges we have to maintain water quality in water bodies in the shift towards a more hydrosocial system. The chapter concludes by reflecting on the nature of water quality measurement and management and the potential for further transitions in relation to its management.

Water supply and sewerage in Waiheke

Waiheke is an island about 40 minutes' ferry ride from central Auckland in New Zealand. While the easterly parts of the island remain rural and are served only by dirt roads, the areas in the west, close to the ferry terminal, form a commuter suburb for Auckland. What is unusual (though not unique, see Gabe *et al.*, 2012; Fam and Lopes, 2015) about the suburban parts of Waiheke compared to other urban built-up and developed areas is that they have no piped water or sewerage. Each house collates its own roof water in a water tank that is then used for all aspects of water supply, often without treatment. Likewise, each house has its own small sewage treatment process that then discharges to the ground.

From widespread anecdotes supported by observations from elsewhere, Waiheke residents are very proud of their own water systems. Many residents claim they prefer to have their own water than any centralised system of supply. One advantage of the system is economy (Gabe *et al.*, 2012): Waiheke residents do not pay for water, though clearly they buy the tank and sewage systems in the first place and also pay for system operation (for example, the use of a pump) and may invest in maintenance. More important is the residents' perception that they have their own water from their own roof and that it is uncontaminated by chemicals or any other external treatment process, and therefore they perceive it as both secure and safe (ibid.; Sofoulis, 2015). As in other locations with rainwater tanks (Woelfle-Erskine, 2015), during periods of drought Waiheke residents manage their own water use to minimise the demand upon their rainwater tanks. In the event of a tank running dry, there is an infrastructure for more water to be delivered by tanker. The island also hosts an array of businesses to support the independent water systems of the island, installing pumps, for example, and cleaning water tanks.

Research suggests that Waiheke residents' perception that their roof water is uncontaminated is erroneous. Simmons and colleagues (2001) collated tap water from households in Waiheke and from three similar non-reticulated 'urbanised' locations in New Zealand. Of 125 tap waters tested, 56 per cent exceeded the local microbiological criteria for maximum acceptable values, with one and two samples respectively containing salmonella and cryptosporidium; they conclude that, together, these findings indicate that the roof water is a potential source of human illness. Their concerns are similar to other technocratic water managers from Australia, who voice worries about the potential health risks arising from the use of rainwater tanks, particularly if they are poorly maintained (Sofoulis, 2015). Compared to many other locations, the long-standing water tank culture of Waiheke ought to mean that the tanks are better cared for than in some locations, as many people have extensive experience of caring for water tanks (Mankad and Gardener, 2016), and community interactions and services enable community learning (Fam and Lopes, 2015).

As well as impacting on the process of water supply, by collecting and storing roof water Waiheke's rain tanks also influence the island's watercourses. Whereas research has enabled us to understand something of the water tanks' influence on water supply processes, there is no equivalent technical research on Waiheke's watercourses that addresses the question: what would Waiheke's watercourses have been like had reticulated water systems existed? Like all areas that have been subject to urbanisation, Waiheke's natural drainage is likely to have become more 'flashy' over the past 50 years and hence also to be more likely to be subject to sudden pollution episodes as water rushes in from roads and other surfaces. Compared to other built-up areas, however, Waiheke will provide less sudden run-off because much of the initial run-off from roofs (though not roads) will be diverted into water tanks. Just as an ecosystem stores water within the soil to slowly infiltrate through groundwater to the watercourse, so water tanks (which hold water back until called upon for toilets, showers and washing machines) slow the rate at which rainfall flows to the island's watercourses, hence smoothing the quantity of flow, and also potentially improving water quality. By reducing some of the flow in the first flush of a rainstorm, there may be some reduction in the erosive effect of rivers and, hence, in their ability to transport sediment. In terms of the impact of the water which is released from household treatment systems, many separate sewage tanks and treatment systems are likely to be functioning less reliably than a large professionally maintained system; however, their impact is dispersed and some further 'natural' cleaning may occur between household treatment and release into a watercourse. The receiving water is the sea off Auckland and the island is not known as a significant polluter of this ocean.

In practice, though the drainage impact of water tanks is not 'known', it is likely that New Zealand water managers would assume that the use of water tanks impacts positively on both water flow rates and quality. In many urban parts of New Zealand, municipal rules require rainwater tanks to be fitted to all new or altered developments (often in additional reticulated water supply), in order to minimise the impact of the increased urban envelope on local watercourses (Gabe *et al.*, 2012). In Australia, water tanks have been widely promoted for their potential impact on water supply, but some managers recognise that they can also have a significant effect on stormwater flows (Sofoulis, 2015).

The Waiheke case demonstrates an interesting contradiction in relation to drinking water quality. Waiheke residents like their own water and perceive that it is safe, but the experts know that it is not reliably safe. This poses a public engagement challenge for the authorities: just telling people that their water may be contaminated is likely to be met with disbelief. A more successful strategy is likely to include recognition and respect for why people like their own water supply alongside the idea that this supply needs to be cared for responsibly.

The Waiheke case is also of broader relevance showing that, in the right hydrological conditions, it is possible to have dispersed water supply and disposal systems in a suburban area – this is contrary to general assumptions. If drinking water quality can be maintained then the system appears to have some strength in terms of minimising development's impact on the water cycle.

In human health terms, it is important that it is roof water rather than groundwater that feeds drinking water sources on the island. Hence, insofar as there is contamination of underground water, there is little likelihood of a pathogen forming a cyclical disease cycle in which illness is reproduced and passed on through drinking water, as would occur were seepage from underground sources used for water supply. In this respect, Waiheke is crucially different from the example of London to be discussed in the following sections.

Chadwick and the vision of a centralised sewer system

In London, at the start of the nineteenth century, most drainage was the responsibility of landowners and parish boards, while foul waste was dealt with on a property-by-property basis through cesspits and night soil men. By the end of the nineteenth century, this situation had utterly transformed: a centralised sewer system collated and treated surface water and foul waste from across the city. This sewered city created in London and a few other cities, including Hamburg and Chicago (Burian *et al.*, 2000), set the model for managing foul waste and drainage in other cities worldwide, even up to the present day. In this context, it is useful to see why it developed the way it did, and hence to consider whether and how the system might have been different. The discussion of London's transition to a 'sewered city' is explored in three steps. The first and, possibly, the biggest step was the shift towards imagining a shared and centralised sewer system. It is this that is addressed in this section, beginning with the preceding conditions, and then laying out key elements in the process of change. The second step concerns the shift in understandings and knowledge about disease: discovery of germs and of waterborne disease vectors created a situation in which individuals could no longer consider themselves safe in their intuitive judgements about water quality. The third step involved the actual construction of the sewer system to serve a set of then separate parishes and districts. These second and third steps are addressed in the subsequent sections.

Like in most urban areas, the antecedents of London's sewerage system are found in the natural streams and watercourses that drained the land on which the developing city was located. Urbanisation brings fundamental changes to such rivers and streams. As an example, the River Fleet in London, which at one time accommodated ships as far as 3 miles north of the Thames at Kentish town, was reduced in flow quantity through diversion for power and water supply through the seventeenth and eighteenth centuries (Tindall, 1980). Though the reduced Fleet still provided fresh water to the rapidly urbanising St Pancras and Kentish Town at the beginning of the nineteenth century, as building spread, the water became 'fetid' and the river was then culverted (ibid.: 26). Pollution arose for two reasons: first, the long-running disposal of waste water and drainage from roofs and roads and, second, increasingly through the nineteenth century, its use to remove foul waste.

With a twenty-first-century mind-set focused on catchments, it is perhaps challenging to understand that, in 1800, the purpose of drainage was to remove surface water from developed areas, with the final destination of the water being a matter of little concern. The overall system was ad hoc, with partial street drainage going into a range of public sewers and watercourses, some of which were open air (Jeffries, 2006). Core responsibility for drainage and any sewers lay with landowners, while natural watercourses, road-paving and drainage of public roads were cared for by numerous locale-specific commissions or boards; for example, nine separate paving boards covered less than a mile's length of the central London road, The Strand (Halliday, 1999). Disputes were resolved and (some) coordination achieved through eight local sewer commissions each covering a different segment of what we now understand as the inner suburbs of London.

The traditional means of handling foul waste was on a house-by-house basis through cesspits that were periodically emptied by night soil men. Though the density of cesspits in urban London must have caused widespread seepage into watercourses, it was not until 1815 that it became officially permitted to put foul waste directly into sewers. The change broke with the long-established norms (discussed in Chapters 3 and 4) in which noxious substances were kept out of watercourses. Halliday (1999) ascribes this change to lobbying by the water companies anxious to make use of the newly fashionable water closet to increase their market for clean water. First patented in 1778, each toilet flush added more than 12 litres of water to less than 1 litre of foul waste (Worsley, 2011). This diluted sewage filled cesspits far more rapidly than previously, increasing the risk of seepage and overflow, and raising the frequency and cost of night soil collection to unacceptable levels. If toilets were to be placed in properties, new means of waste water disposal were needed. Also, flashy and dirty surface water sewers were thought improved through the regular flushing arising from a (relatively constant) supply of foul sewage.

Pressure for foul waste disposal in middle-class areas may have created the regulatory context for the spread of sewers; but further shifts in thinking were needed before such sewers were retrofitted systematically into the existing poorer suburbs. Two important factors underlie the shift towards thinking about a sewage *system* for London. First, in nineteenth-century thought, smell was the cause of disease and disease was the cause of poverty. Specifically, major epidemics such as cholera, typhus and typhoid were thought to flourish in filthy environments and to be transmitted through foul-smelling contaminated air, or 'miasmas'. By addressing smell, it was suggested, disease-driven poverty would cease, and a huge burden of caring for poor could be relieved. Hence, sanitary reform, which included the building of sewers as well as the clearance of slums, became a moral imperative for many (Allen, 2008).

Second, poor neighbourhoods had become significantly more crowded during the early nineteenth century, with the London population doubling between 1801–41 (Jeffries, 2006: 2). As Johnson (2008: 115, emphasis original) delicately words it, this raised the pressing question: '*What are we going to do with all of this shit?*' The export of cesspit contents to farmland outside London became relatively more expensive as the city grew and the farmland was more distant. The problem was further exacerbated from 1847, when the import of cheap Guano (bird droppings) competed with human waste as a source of agricultural fertiliser. As the rural market for night soil declined in value, the cost burden of cesspit emptying for homeowners and landlords further increased (Jeffries, 2006: 2), causing more and more localities in which exposed sewage festered *in situ*. The impact of this on even respectable London neighbourhoods is hard for us to imagine from the twenty-first century. As the account of the neighbourhood around Broad Street in the next section will illustrate, everywhere must have smelt, and for all apart from the wealthiest, other people's sewage was encountered on a daily basis.

The principal champion in London's switch to imagining a centralised sewage system was Edwin Chadwick, a central and early promoter of sanitary reform. Initially on the Poor Law Commission, subsequently as a Commissioner of Sewers, and latterly as Chair of the Board of Health (Halliday, 1999), Chadwick's key motivation seems to have been his belief that, by removing filth, disease and hence poverty could be relieved. He argued that this was economically sound for two reasons: first, pressure on the poor law would ease if poor people did not get sick; second, the sewage itself could be made to make money. In a vision that Edwin Chadwick shared with many other contemporary commentators, the clearance of cesspits and local accumulations of sewage was the first step in a vast future, economically self-sustaining and circular sewage management system in which human waste was an economic resource collated from the population and piped to farmers to fertilise fields.

In the 1840s Chadwick criticised the seven sewer commissions (the aforementioned bodies which coordinated drainage across London) for being ineffective, corrupt and standing in the way of a coordinated drainage system. The sewer commissions' original role of arbitrating in drainage disputes between parties had been significantly developed, and from about 1800 they took the initiative and began to construct, repair and cleanse public sewers, and also to enforce regulations to ensure that householders' drains were built to a sufficient standard (Sunderland, 1999). As described below, the differences between Chadwick and the sewer commissions related to many technical details, but came down to Chadwick's focus on clearing the administrative and physical obstacles to making an interconnected London drainage system a reality. For Hamlin (1992) and Sunderland (1999), the critiques Chadwick made of the commissions were unfair. Sunderland, for example, argues that the commissions were 'efficient, honest, innovative and receptive to the needs of their ratepayers' (ibid.: 372) and that, despite their many constraints, they 'kept pace with the ever-changing sanitary needs of the capital and built a sewer system far more advanced than that of any other urban area either in Britain or abroad'.

Nevertheless, in 1848 Chadwick was successful in causing their abolition and replacement by a single sewer commission for London. In parallel with this development, the Nuisance Act compelled landlords and homeowners to connect their properties to a sewer where there was one nearby. The new Metropolitan Commission for Sewers embarked on an energetic programme of draining cesspits, and as much as 61 million litres of sewage had been transferred to the Thames by the winter of 1849.

The process of sewer construction did not meet with universal approval. One significant objection was that it involved an unwarranted degree of governmental interference in people's daily lives. There was also a particular middle-class horror of being forced to open a home up to physical connection with the dirty and diseased working-class areas (Allen, 2008). Moreover, Chadwick's argument that the loss of an individual's labour through disease damaged the whole economy ran counter to laissez faire norms of the time, in which health was an individual concern (Lee, 1994: 283). A second issue arose because the changes required large expense, and coordination across London by a large administration rather than management by local Boards of Work (Allen, 2008: 40). In this respect it involved 'big government' that worked against accepted principles of free trade and local administration (or 'subsidiarity' to the smallest appropriate level of government, as it would be termed in modern parlance) (Johnson, 2008: 113). A third concern was environmental: by clearing the cesspits and constructing sewers in towns, many small pollution problems were transferred into a smaller number of large problems – namely, the quality of rivers, and not least the Thames itself.

It now seems clear that, in many respects, Chadwick's vision was driven more by ideology than practical reality. Chadwick and his allies presented themselves as at the forefront of a new technology opposed by backward-looking engineers employed by vestries and by the sewer commissions. Yet, according to Hamlin (1992), the 'opposition' was far from united, but rather concerned with real practical difficulties of the proposed schemes. Chadwick gave way to some of the practical points made by these opponents; the transfer of sewage to the countryside was postponed to an imaginary future, it was also conceded that any sewer pipes constructed needed to carry storm drainage as well as sewage – though the space needed for the latter remained minimal in his view. A key enduring difference between the parties concerned the size – and hence the cost – of said sewer pipes. Chadwick advocated small clay pipe sewers in all locations; in contrast, the group Hamlin calls the 'engineers' suggested larger, brick-lined versions, varying in size and location according to anticipated quantities of sewage and surface runoff. Chadwick believed that the sewer commissions and local councils were buying in overly expensive sewer solutions from engineers who were just ensuring that they had a substantial construction job to carry out. Table 6.1 summarises some of the technical factors that played into these arguments. Chadwick's arguments are largely based on what we now see to be mistaken hydrological assumptions about the extent to which small pipe flow is able to increase velocity if there is more water upstream.

TABLE 6.1 Arguments between Chadwick and the engineers in relation to the size of sewers

Issue	*Chadwick's argument for small cheaper clay pipes*	*Engineers' arguments for larger brick-lined sewers*
Sewage volume	Small pipes can carry small and large volumes	Large pipes are needed to carry large future demand
Storm flow	Storm flow can be accommodated through faster flow in pipes	Large pipes are needed to accommodate the flow from London expanding
Cleaning/ blockages	High (and preferably constant) velocity minimises blockages and prevents a need for cleaning	A man has to be able to enter the sewer to clean blockages that might arise
Inappropriate material	Inappropriate material should not be put in the sewer	It is not possible to stop people putting inappropriate material in the sewer

Source: Material drawn from Hamlin, 1992, and Sunderland, 1999

These differences came to a head towards the end of 1852. Croydon's newly constructed sewer system had followed Chadwick's design requirements, but only weeks after its conclusion, breakages, blockages and fever occurred (Hamlin, 1992). An initial inquiry by Chadwick's Board of Health blamed poor contractors for the failures. A follow-up Home Office inquiry found the opposite, however. It blamed Chadwick's Board of Health for being blind to its own responsibility for having inspected and approved the design. Meanwhile, Bazalgette, Chief Engineer for the Metropolitan Board of Sewers, inspected other Board of Health-approved sewers and reported numerous difficulties where he found deposits, cracks and breaks. Mounting opposition to Chadwick argued that the General Board of Health had done damage by trying to suggest a single solution as a panacea when the nature of sewerage required variations in levels, in sizes of sewers and in the nature of work. These difficulties led to Chadwick's expulsion from the Board of Health in 1854.

This historic case provides a useful example of the contradictions of a 'champion' for a transition. Supported by the precursor decision to let sewage be released into surface water sewers, Chadwick's crucial contribution is that, even in the face of the free-market, small-government preferences of the day, and in the context of the confused and overlapping jurisdictions of Victorian London's urban infrastructure, he had the chutzpah to imagine that a joined-up single London sewer system could be created and administered. He was also a pioneer in daring to think that there should be collective state responsibility for issues relating to health or the environment. Notwithstanding animosity between him and Bazalgette over details of sewer design, Chadwick's work can be seen as a crucial precursor to the process of sewer development that Bazalgette would later lead and be championed for. Despite these great strengths, Chadwick might also be argued to have done significant harm. He destroyed the effective sewer commissions in fighting for his idealist vision of a centralised sewer system. He insisted on inadequate sewer designs, with the effect of great expense and disease. Moreover (as will be further discussed in the following section), his priority on emptying London cesspits almost certainly increased the rate of spread of infectious disease through the 1850s.

From a current perspective, the enormous change in thinking recounted above was the shift from the idea that foul waste management was an individual problem to a collective one. Albeit driven by notions of miasma rather than germs, this can be seen as the foundation for an understanding of public health as a collective concern. A concomitant impact was that it framed sewage as a liquid waste and distanced the individual from taking responsibility for the nature or content of their liquid waste and its impact on others.

It is interesting how the changes described in this section structure the categories we have today. Flush toilets might have been rejected for causing unacceptable foul waste disposal problems, but instead they were welcomed. Likewise, Chadwick was ahead of his time in seeking a separate foul and surface water waste system, and though systems were indeed divided in this way from the 1950s onwards, the legacy of many combined systems causes expense and pollution to this day. Moreover, the large sewers implemented against Chadwick's advice removed any need for public education about waste water disposal, a factor that may be seen as a precursor to the contemporary issues water companies face with fat, oil and grease ('FOG') and sanitary products in sewers. Chadwick also had one more as yet unrealised aspiration – that the foul waste should systematically be used as a source of fertiliser. Who knows – given better technology, oil shortages and more sewer separation, perhaps this vision too will be realised in the future.

John Snow and the Broad Street pump incident

The story of the Broad Street pump cholera outbreak provides a detailed window into water and sanitation within a small part of London in 1854. Though it post-dates the requirement that cesspits were connected to sewers, it shows the limited effects that this change had wrought. It also illustrates the nineteenth-century reaction to epidemic disease, and the challenge faced by science that does not accord with accepted understandings. In terms of this chapter's themes about centralised and decentralised systems and responsibilities, the story details the emerging recognition that individual sensory perception is insufficient to judge the safety of drinking water. In this respect, we can see the knowledge processes making a real contribution to driving the centralisation of water supply management.

In his account of the Broad Street pump incident, Johnson (2008) describes this London neighbourhood's population as 'a mix of the working poor and the entrepreneurial middle class' (ibid.: 19), with residents including tailors, shoemakers, domestic servants, masons and shopkeepers. There were a large number of pumps within this crowded part of London, so most people had to travel less than 200 yards (188m) to fetch water. It is notable that water was trusted to the extent that people regularly drank cold water directly from the pumps. Different pumps had different reputations and the deep Broad Street pump was known for its purity, visible clarity and pleasant taste. In addition to the public pumps, the adjacent workhouse had its own well, while the beer factory was served by the New River company. In terms of waste water, the experiences of the multiple residents at a particular critical property at 40 Broad Street depended on the location of their rooms in the building. Those located at the back of the building threw their waste water out of the windows into the backyard, which could only be traversed dryly by utilising carefully placed bricks. Meanwhile, those located at the front of the building disposed of waste water in a cesspit, which went into an overflowing cellar (later discovered to be located just 2 yards from the Broad Street pump's shaft, and the source of initial contamination to the latter). Though, following the 1848 legislation, there was a connection from the cesspit to the local sewer, it was blocked and the material was effectively dammed such that it sat in the cellar. Chadwick's study of London housing in 1849 suggests these unsanitary conditions at 40 Broad Street were not uncommon; of 15,000 homes, 5 per cent had human waste piling up in the cellar (ibid.: 115).

The Broad Street cholera outbreak was distinguished from others by its intensity – a person could move from healthiness to death in just 12 hours – and its relatively narrow geographical scope, centred around the Soho part of London. With the emergence of the epidemic, many people fled the area. As with previous epidemics, the means through which the cholera transmitted itself was a matter of debate. Local rumour suggested that it might have arisen from local plague pits, revealed through recent excavations for new sewers. The official sanitary authorities signalled their understanding that contagion was transmitted through the air because they soaked the local streets in chloride of lime as a means to clear the problems (ibid.: 112). Individually, locally resident doctor John Snow and clergyman Henry Whitehead began investigations into the outbreak that were then brought together in the official vestry study of the event. John Snow had the advantage of having already studied the 1848 outbreak in detail, and had developed the theory that transmission was achieved through water. He had been part way through an arduous process of retrospectively tracing the source of victims' drinking water in order to explore his thesis that that 1848 epidemic was waterborne. The arrival of a new epidemic, just streets from where he lived, provided an opportunity to study cholera as it was happening. He brought the methods he had developed in relation to the 1848 outbreak to bear on events in 1853.

What is now seen as a close epidemiological analysis was carried out on all cases of cholera in the 1853 outbreak, tying Snow's theories about waterborne transmission together with Whitehead's detailed knowledge about the sources of water and habits of individuals in the vicinity. Their analysis identified the Broad Street pump as the cause of transmission, and an infant at 40 Broad Street as the crucial 'index' case. Though the work can now be seen as providing absolutely convincing arguments for the waterborne transmission of cholera, the London sanitary authorities repudiated the vestry's conclusions about the cause of the Broad Street outbreak. Their claim, maintained for over a decade after the outbreak, was that miasma from the obvious smells in the area was the cause of disease. Their view that Snow was blinded by his theoretical starting point (which is what we might now say about the sanitary authorities) demonstrates the difficulties faced by those seeking to question established ways of thinking and acting.

The next significant London cholera outbreak occurred in 1866, centred on East London. Though Snow was dead by this time, William Farr, a medical statistician who was investigating the episode linked up with Henry Whitehead and reproduced Snow's methods from Broad Street in relation to the new outbreak. This study revealed the East London Water Company's water supply as the source of the epidemic. The not-yet-connected local sewer network, a water intake downstream of old sewage outlets and poor company filtration mechanisms had combined to expose Londoners to one final cholera episode. Farr's investigation of the epidemic brought renewed attention to John Snow's preceding study of the Broad Street outbreak; Farr's work helped to establish the germ theory of disease and prove the waterborne nature of cholera transmission (ibid.: 211).

The story of the Broad Street pump incident shows how people's understanding about the transmission of disease framed (or, rather, did *not* frame) the way that water quality was measured. Rather like the residents of Waiheke today, for the dwellers of Broad Street smell, colour, taste and previous experience denoted perceived quality. Indeed, perhaps we should sympathise with their perception: it is odd that the abominable smell should be harmless whereas water that appeared clean should actually kill. Even for Snow – whose suspicions were focused on the water – testing was confined to basic chemistry such as the measurement of pH. In practice, the samples provided no evidence that was useful in the context of the testing and knowledge regimes of the day. Rather, the circumstantial information arising from Snow and Whitehead's epidemiological study offered a new type of evidence not previously known, and in contradiction to contemporary understandings. In this context – and however compelling the case made – it is perhaps not surprising that it was another 12 years before their theories were more broadly embraced and understood.

While the story of Chadwick in the preceding section considered how more centralised sewage systems came about, the story of the Broad Street pump illustrates arguments for centralised delivery of and responsibility for water systems. At least in the mid-Victorian context, only larger companies could ensure knowledge of, treatment of and testing of water quality. In the context of the 1866 discovery about the waterborne transmission of cholera, reliance on local street pumps for the provision of potable water continued to decline in favour of larger commercial water companies' piped supplies.

Bazalgette and the construction of interceptor sewers

If Chadwick introduced the idea of a single connected sewer system and state responsibility for public hygiene and health, Bazalgette, to a significant extent, put it into practice. Bazalgette was chief engineer at the Metropolitan Board of Works. Whereas Chadwick's motivation was an ideological one to address poverty,

Bazalgette was an engineer, interested in meeting his employers' requirement for a solution to the problem of foul waste in London. The story told in this section is essentially a problem about centralised governance: how did Bazalgette, and others committed to developing a centralised sewerage system, achieve the political and financial buy-in to support a large and integrated project of 'interceptor sewers' to keep London's sewage out of the Thames?

The task at hand was connecting all of the various private and public sewers previously constructed by the separate Metropolitan Sewer Boards before 1848 to make a single sewer system that would work and that could be maintained. As Bazalgette himself commented in 1852, the construction of local sewers to relieve smells from inner-city suburbs was the initial priority (Halliday, 1999: 54–5). With a miasmic worldview, this has some sense. The emphasis was on removing danger from the very densely populated inner suburbs. If the tradespeople working by an already smelly Thames had slightly more smells, it was a cost worth the benefit of many smell-relieved city dwellers; besides which this cost was for the short term only. In the longer term, the Board would address the more substantial and more difficult task, to develop plans for large 'interceptor' sewers to take sewage downstream from London sufficiently that it could be released without the tide carrying it back into the city. These sewers were so called because they intercepted the smaller sewers, collecting their contents before they disgorged into the Thames. The construction of such interceptor sewers was both a physical and political challenge.

The construction of interceptor sewers was a physical challenge because of the project's sheer enormity. Not only were many existing properties along the bank of the Thames to be removed, but the 'bank' of the Thames was to be extended into its former channel, changing the appearance of significant parts of central London. A particular technical issue arose because of the Thames' tidal nature. In order that gravity-driven sewage outfalls flowed into the river, they frequently entered the river at a level below the high-tide mark. The effect of this was that sewage backed up in the sewers when the tide was high and then disgorged into the river when the tide was low. In dry weather, a high proportion low-tide flow was sewage. Interceptor sewers along the banks of the Thames would do something to transport this problem downstream, but they would not prevent it. However far downstream the outfall was placed, sewage might still enter the river at low tide. Moreover, as the tide turned, the sewage that one might have hoped would have gone out to sea could be swept back up the Thames into London, where some of the water companies' intake pipes were located.

In creating a single 'Metropolitan Commission for Sewers', Chadwick's 1848 reforms had done something to overcome the political challenge arising from London's medieval system of governance. For sewers (but for sewers only) there was one governance organisation covering the whole of the then built-up area. But the Commission still needed to grapple with the needs of different parts of London, as well as being attentive to those living downstream and outside their boundaries. Part of this dilemma related to cost: should Londoners pay themselves for the cost of this system, or should the huge expense of conveying sewage away from London be shared by the country as a whole?

During the decade 1848–58 a series of Metropolitan Commissions for Sewers (and subsequently the Metropolitan Board of Works) struggled to resolve these physical and political issues. Technically significant progress was made in the completion of a survey of the existing sewers. A public competition for ideas, and then iterative processes of planning and replanning led to successive detailed plans for intercepting sewers. Political progress was also made when the new Metropolitan Board of Works started to include representatives from different parts of London. The most enduring differences, however, brought technical and political issues together in disputes over the location of the system outfalls and the scheme costs. Central government insisted upon the highest environmental standard, hence with very distant outfalls far downstream from London. Meanwhile, the Metropolitan Board of Works sought to minimise costs to ratepayers and argued that the outfall could be extended from the original plans, but that this would require funding from central government. After several years of stand-off, the need for a resolution to this issue was amply demonstrated to government in 1858 when the long, hot summer meant that the stench from the Thames, which – thanks to sewer construction of the previous decade – was now a flowing cesspit, made parts of the Houses of Parliament unusable (Halliday, 1999). Resolution took the form of a new act of parliament, passed in an exceptionally short 18 days, that allowed the Board loans guaranteed by government and accompanied by an easement of environment controls in favour of a 'practicable' solution.

The period 1858–75 marked the grand period of sewer construction, with the official opening of the system in 1865, and further openings in 1867 and 1875. Although it was a very large construction project, which caused significant disruption as well as being at great cost to current and future ratepayers, Bazalgette and the Metropolitan Board were remarkably successful at keeping the UK media and government onside during this process. There are a number of factors that might be said to have contributed to the project's 'success'. First, because action did not start until the problems of the contemporary system had been so explicitly demonstrated, interceptor sewer construction began in an atmosphere of widespread agreement about the need for action. Second, the Board and Bazalgette seem to have been adept in media management, inviting journalists on visits to view the construction and keeping up a flow of information to the newspapers (ibid.). Third, Bazalgette pioneered a new type of engineering contract, with clear specifications and penalty clauses; he also clearly put enormous personal effort into the specification and management of the works (Hughes, 2013).

The fudge that was the 'practicable' 1858 resolution of the impasse over the sewer outfalls came back to haunt the Board a decade later. Much greater residential development in the lower Thames during the intervening decade made the proposed location for the raw sewage outfall highly problematic. Eventually, settling basins enabled the relatively clean water to be extracted and the solids collated for transport and dumping out at sea (a system that continued to be used until 1998).

Hughes (ibid.: 689) argues that: 'The London sanitation projects were perceived at the time as being successful and with the benefit of hindsight they still appear successful.' Without completely disagreeing with this claim, it can nevertheless be pointed out that Bazalgette's interceptor sewers were very large. They were successful not only in serving the calculated increase in London population, but also in remaining the mainstay of London's sewer system to this day. Is it fair and right that the ratepayers of the late nineteenth century should have made such a significant expenditure and investment and that we are still enjoying its fruits? A modern 'adaptive' approach might seek to find a more scale-able solution, with more moderate contemporary investment accompanied by options for subsequent future development.

The development of London's interceptor sewers can be seen as a desperate and expensive solution to an accumulated set of problems caused by urbanisation. An important long-term precursor was the encouragement of toilet technologies through the decision to allow foul sewage in the drainage system, driving the subsequent huge increase in water closet installation and hence in the quantity (and liquid quality) of foul waste. A more immediate driver was the problem of smell in residential areas that was believed to be the cause of disease, such as the cholera outbreak in Broad Street. Ironically, by taking action to solve the many local 'smell' problems in London another larger problem was caused; the confluence of many sewers upon the Thames resulted in the 'big stink' – and perceived disease threat – along its course. Collective action to build the interceptor sewers was carried out by central government because the smell of the river was seen as bringing shame on the city. The great stink therefore offered a strong demonstration of 'need', hence helping to carry public support for an infrastructure investment programme of unprecedented size.

The construction of the interceptor sewers can be seen as the full implementation of the 'centralisation' of foul waste management that was begun under Chadwick. Perhaps the most remarkable thing about the story of London's sewers is that they achieved what they aimed to – relieving London from epidemic disease – but for the wrong reason. The entire process of campaigning for and developing the sewer system was premised on a miasmic theory of disease transmission. In the long run, by removing smells from London streets the sources of water contamination were also removed. However, the short-run priority focused on building local sewers to flow into the Thames – a source of drinking water – is likely to have significantly increased the transmission of waterborne diseases in the period 1848–65 as more potentially contaminated outflows overlapped with points where water for drinking was abstracted.

The perceived success of the London sewer system caused its model to be repeated in many cities around the world. Some of the water issues we deal with in cities today can be seen as the legacy of these decisions. An example of these contemporary water issues is demonstrated through Bradford Beck.

Bradford Beck: a contemporary urban watercourse

Bradford Beck is an urban stream draining about 60km^2 around the intensely urbanised post-industrial city of Bradford. 'Beck' is a northern English word for 'stream' and, before coal mining and textile manufacture changed 'Bradford Dale' (valley), Bradford Beck was a fast-flowing watercourse serving a hilly area in the eastern Yorkshire Dales, and draining into the River Aire at Shipley. In its current form, some tributaries (becks) retain rural headwaters, hinting at how the catchment of the entire Bradford Becks system used to be. Many other tributaries have entirely urban catchments, and are canalised or culverted (covered) for their entire course.

Since 2011, the Aire Rivers Trust and its daughter community group – the Friends of Bradford Becks (FOBB) – have been seeking to improve the beck and its tributaries, hoping both to address water quality issues and also transform the aesthetic value of the beck for the city of Bradford. This activity means that there is information about the state of the beck and that efforts are being made to effect change. Bradford Beck and its sub-catchment (henceforth 'Bradford Becks' or the becks) is examined here because it provides an exemplar of the difficult problems we have managing water quality in watercourses within urban areas today.

Bradford Becks count as 'heavily modified watercourses' in EU Water Framework Directive parlance. Victorian canalisation and culverting mean that the physical courses of the becks have, indeed, been 'heavily modified' for many years. Such straightened river courses are problematic for fish and other wildlife because there are few pools between the fast-flowing water, limited gravel deposits in which insects or other species can breed and little of the variation in light and warmth associated with overhanging vegetation in a 'natural' stream. The catchment is also modified due to urbanisation. This reduces overall flows within the beck because much surface drainage enters the combined sewer system; however, because urbanisation keeps water on the surface and speeds up its flow, those waters that do enter the becks rise rapidly during a storm. In the 1960s, the consequent flooding problems led to the construction of a large flood overflow channel that bypasses Bradford city centre, taking high flows during wet weather conditions.

In addition to these physical modifications to the channel and quantities of flow, the quality of the beck's water are also modified, in that rainfall is vulnerable to chemical modification through pollution. Potential sources of pollution include 51 combined sewer overflows, misconnections from domestic and industrial sewage (misconnections occur when a pipe that should go to the foul sewer is connected to the surface water sewer in error), leachate draining from historic landfill sites and surface water connections carrying road dirt. Of course, these potential sources of pollution are nothing compared to what the same beck has carried in the past: through much of the Victorian era the becks were an open sewer, and as late as 1993 significant industrial pollution systematically polluted one of the beck's tributaries. This long legacy of pollution means that the becks – or at least their central and most polluted sections – are less likely to have significant reservoirs of species diversity to recolonise their waters.

If there are challenges in deciding how we should define clean drinking water, the conceptual and practical difficulties that pertain to assessing the quality of rivers are even greater. Between 1990 and 2009 the UK Environment Agency used a measure of river water quality based on chemical contents of samples called the General

Quality Assessment (GQA). The story the overall GQA figures tell is of a massive improvement in UK river quality over this period. Bradford Beck is no exception to this trend, with the river changing from 'poor' to 'good', with the closure of manufacturing plants that had polluted the beck in the early 1990s marked by very notable reductions in the ammonia content of the water. The focus on chemical standards demonstrates a concern with addressing point sources of pollution – that is, sewage outfalls or industrial works that had a distinct chemical impact on water quality.

The EU Water Framework has brought in a more complex measuring regime based on both chemical and ecological objectives set for different rivers and other water bodies. The standards were designed to support the Water Framework Directive's aim of stopping the deterioration of water bodies and supporting their improvement towards 'good ecological status'. The classification involves assigning waters to one of five ecological statuses: high, good, moderate, poor or bad. The method of assessment is based on a 'one out–all out' system, in other words, if the becks score badly for one of the parameters assessed then they are rated as 'bad' overall. In practice, the assessment of Bradford Becks in 2009 rated the becks as 'bad' because there was a very poor macro-invertebrate count. Nevertheless, the assessment indicated that there were some fish present, and also that the chemical quality of the water is moderate (based on moderate levels of ammonia).

The FOBB conducted their own investigation of water quality with the aim of updating the EA information, but more particularly to diagnose why the water quality was poor and to identify where the sources of pollution were located in order that the problems could be addressed. To this end, they recruited volunteers to carry out sampling across 19 sites, repeating the sampling on different dates and at different times during 2011–12. In addition, a visit from a fish expert helped the group to explore aspects of the beck that were – or had the potential to be – friendlier to fish. Practical issues prevented the group repeating the macro-invertebrate study that formed the basis of the EA's 'poor' designation.

The group's study supported the continued designation of the beck as 'poor' quality. The problematic findings are represented in Table 6.2. Effectively, the work showed that there were some issues with effluent pollution throughout its length, as demonstrated by the ratings in terms of biological oxygen demand. It also demonstrated a series of site-specific problems, relating to various potential misconnections, combined sewer overflows or sources of industrial effluent. The group's findings are presented on their website, alongside a record of specific pollution anecdotes spotted by group members and supported by photographic evidence. Though the latter is unlikely to be complete (i.e. there are likely to be many pollution incidents that occurred when nobody was looking), the anecdotes serve as a useful reminder that systematic sampling such as that carried out by the group cannot cover occasional incidents.

The Environment Agency's designation of the beck as poor occurred because of low levels of macro-invertebrates. These include aquatic insects, snails, crayfish and worms and can be seen as a good indicator of a healthy stream because of their high numbers, known pollution tolerances, limited mobility and dependence on the land

TABLE 6.2 Problematic findings from study of pollutants in Bradford Beck*

Parameter	*Indicates*	*Occurrence of problems*
Biochemical oxygen demand	Domestic or industrial effluent	All sites, all occasions
Chromium	Presence in geology, could be carcinogenic	At around threshold concentration at most sites
Ammonia	Human or animal wastes and landfill leachate	Specific sites and diluted downstream
Zinc and iron	Industrial waste or leaching from contaminated land	Specific sites and diluted downstream
Caffeine	Human sewage through misconnections	Specific sites (mostly to the east of the city)
Chloride	Domestic or industrial effluent	Specific sites and diluted downstream
Boron	Detergent	Specific sites and diluted downstream

Source: Adapted from Aire Rivers Trust, 2012
Note: * In addition to the parameters shown, other parameters tested and proving unproblematic included phospherous, tetrachloroethylene (used in dry cleaning), road-related compounds, organic quality and other heavy metals including arsenic, cadmium, copper, lead and nickel

around the stream. However, as evidenced from Bradford Beck, they do not indicate the existence of fish life. As will be discussed in Chapter 7, they also are not necessarily a good indication of what is valued by local people. Practical issues meant that FOBB did not sample macro-invertebrates. It is possible that the pollution recorded in their study contributed to the low macro-invertebrate life in the beck. However, other factors that may be equally or more important are the lack of variation in the flow and lack of grit or other varied sediment material in which to lay eggs.

The different ways of assessing the quality of Bradford Beck highlight a philosophical question over how river quality should be measured. The real question that lies behind these different ways of measuring water quality is 'What do we want from our rivers?' If we are focusing on the chemical quality of the water, then we are implying that we want the water to be free of contamination and, hence, potentially available for water supply, or accessible for swimmers or surfers. If we focus on biological quality, then we are looking for something else: a complex, functioning and interacting ecosystem can be 'artificially' created where it has not been before (e.g. in a newly constructed pond) or where it has been long absent (e.g. in Bradford Beck) – but it is not a simple process. And we might, additionally or instead, be more concerned with aesthetic quality, which will be explored more in Chapter 9.

There is then the question of what to do with rivers that are labelled 'poor'. Not all such designations lead to action. The use of blue flags to designate clean bathing waters has been a significant public communication success through which action on the quality of bathing waters has been promoted through the UK and elsewhere. Were Bradford Beck to flow into an area with designated bathing waters, then its quality might be an issue of greater official concern. However, it was industrial decline rather than the poor chemical quality of the beck that prompted changes in its chemical status in the early 1990s; and, likewise, the Water Framework Designation of 'poor' has had little impact. Rather, action on Bradford Beck relies on its 'luck' in having an action group of concerned residents seeking to address its quality and accessibility.

Following their studies of the beck's context, the Aire Rivers Trust produced a catchment management plan for the beck. This includes a range of 'do-able' actions, including a section of river renaturalisation in the lower part of the beck, following up the water quality monitoring with the identification of specific pollution sources, removal of 'contaminated water' signs that lined part of the beck's banks and the further development of sustainable drainage within the context of the council's flood risk management strategy. The extent of possible activity is limited, however, by the group's status as a set of volunteers who have come together to campaign for the improvement of the river and its tributaries. The group may have numbers and enthusiasm, but they have neither money, nor control, nor official responsibility for the becks.

In seeking to influence the state of the beck, the Aire Rivers Trust needs to achieve cooperation from a range of relevant individuals or organisations, all of which make their own contribution to the beck's governance. These include the riparian owners, the local council, the water and sewerage provider and the water regulator. Riparian owners are the people who own plots of land or buildings on each side of the beck, or bridging its culverts. Legally they have responsibility for preventing blockages or pollution for their section of the beck. However, and not surprisingly, individual householders or small businesses are often unaware of these 'responsibilities' and have neither the knowledge nor capacity to carry them out. The local authority, Bradford Metropolitan District Council, is directly responsible for caring for the beck and the surface water sewers that drain into it. Relevant departments include the drainage function, the highways department, parks and nature conservation as well as the planning department. One specifically designated role for the council is as the 'Lead Local Flood Authority', with responsibility for coordinating the prevention and mitigation of floods. The local water and sewerage provider is Yorkshire Water, which is responsible for the engineered drainage system, often working in parallel to the beck, that causes combined sewer overflows. Meanwhile, the regulator, the Environment Agency, has oversight of the Water Framework Directive and its implementation, with an interest in the beck and responsibility for monitoring it, but with limited direct levers to ensure that others take action.

The case of Bradford Beck can be seen as a not untypical example of an urban watercourse: continuing to exist, diminished in size, partly buried, largely forgotten and (until recently) little loved. The fundamental issue concerning the beck is 'What do we want from it?' The existing management regime is largely focused on ensuring that it does not cause too much additional pollution in the River Aire. The Aire Rivers Trust has dared to challenge this approach, suggesting that the beck could be a thriving urban watercourse with a healthy ecology, and an asset to the city.

Its ongoing challenge (discussed more in Chapter 9) is to create sufficient understanding of and belief in the possibility of its vision to drive a set of institutional partners from lethargy to action.

Concluding qualities

This account has ranged in time and space. What are the overarching lessons about water quality management in a hydrosocial system?

The chapter has demonstrated the strange and almost nebulous nature of 'water quality': we want water to be 'right' for the purposes to which we put it, but we don't quite know what that means or precisely how to assess it. One of the difficulties has been the knowledge and resources available to assess water. Like the former residents of Broad Street, residents in Waiheke judge their water by smell, taste and experience. Likewise, Victorian residents of London and the FOBB today judge the quality of their watercourse first and foremost by appearance and smell. Even when advances in knowledge mean that contemporary residents of Waiheke and Bradford technically have more tests carried out on their waters, as we have seen, they do so infrequently because the costs and practical challenges of reproducing expert tests are significant. At the heart of the choice not to test lies the question 'What aspects of quality really matter?' Is the existence of pathogens in a Waiheke resident's rain tank a problem if they do not make the resident ill? Likewise, if there are fish, and if the river becomes a physically more pleasant environment to visit, do the macroinvertebrates that drive the Environment Agency's assessment of Bradford Beck really matter? And, finally, though importantly, in a not-quite-tangible way, the waters we are discussing are flows. Consequently, any measurement or assessment of them is a snapshot recording a moment in time and not an absolute state; more information can always be found with more snapshots, but what is the benefit? Depending on what we want from our waters, we select sampling and testing regimes to help make our arguments or to help achieve the changes we want.

These reflections suggest that new measurement and assessment technologies have the potential to have a transformative effect on water management. The expertise of scientists may have contributed to the centralisation of water management processes over the nineteenth and twentieth centuries, but new means of measurement have the potential to democratise water management and, hence, to contribute to reconnecting people with water. If available to the residents of Waiheke or Bradford, easy means of measuring water quality would provide the information with which to engage and consider whether new and different qualities are desirable. Where they were desired, frequent testing and assessments of waters could be carried out, enabling more nuanced and localised judgements to be made. For example, Waiheke residents might conclude, 'this water does not upset my family but it does my guests', while FOBB volunteers might conclude 'the pollution must arise from the combined sewer overflow at x location'.

Closely connected to the issue of measurement, the chapter throws light on contemporary mechanisms for managing quality. The historical story of water quality is closely tied to that of urbanisation: our changes to the landscape and the imposition of our wastes upon it impact upon the qualities of different water bodies. Where water used to be drinkable it became less so, and where aquatic life used to thrive it thrived less, or disappeared altogether. To these problems – that occur whenever a city grows – a particular set of historic circumstances caused the development of a new system of disposing of human and other liquid waste through sewers. In terms of the historical account of London, without the water companies pressing to grow their market via the water closet our sewers might have been reserved for surface water. The 'problem of poo' might, therefore, have continued to be seen as a solid waste issue; alternatively Chadwick's initial vision of a separate surface and waste water system might have been realised, and human waste might have been harvested for fertiliser. Likewise, without the big stink, another decade might have passed without coordinated investment during which different solutions to London's waste problem might have emerged. These historical 'might haves' are important to remember. The capital city of the (then) most advanced country developed its sewer system as it did because of a series of historical accidents; if it had developed differently, then models of water and waste water management across the world might have been different too. The growth of technocratic water systems that shapes our contemporary water experience *might not* have happened, or might have happened in a different way.

The historical story also shows an evolution in pollution detection and treatment processes. London's story shows how the interconnected nature of the water cycle drove multiple processes of centralisation. Institutionally, what we can observe in the period 1845–70 is the development of sewers providing a step towards a professionalised city government. The perceived need for more collective action drove a process of centralisation, while the extent of expenditure required professional project managers like Bazalgette to oversee and manage the work. It should be noted, however, that the perceived need for coordination was confined to the management of the underground sewer network, while water supply continued to be delivered by private companies, and the management of surface water drainage and any non-sewered watercourses fell to the vestry or local authority: a functional separation of the governance of these different elements of the water cycle that remains in England to this day. In the context of contemporary calls for the integrated management of the water cycle, this legacy creates special challenges for the English. More broadly and internationally, the example highlights the long legacy issues that arise with investments in water management.

From the 1860s onwards, the 'success' of Bazalgette's sewers in London, combined with its place as the power-house of industrial development, was an important factor in the export of the same system all over the world, propagating both its advantages – as a means for managing public health – and disadvantages – namely the pollution issues and highly modified local streams with which it is associated. London can, therefore, be seen as one of the key demonstrator cases providing the root to today's assumption that cities have centralised sewer systems, making exceptional cases like Waiheke so interesting. Here, an urban community thrives with no external water supply and no external sewage treatment. The example demonstrates that, in some circumstances, our assumption that urban areas need large centralised systems is wrong. Meanwhile, within the sewered cities, the state of Bradford Beck and many other urban streams can be seen as another part of Chadwick's and Bazalgette's legacies. Denuded of regular flow, culverted and forgotten, apart from when they smell or flood, these urban watercourses have been – and largely remain – unloved.

In an imaginary world in which such legacies could be lightly discarded, how would we like to manage water? Without seeking to draw up a blueprint for future water management, our contemporary understandings, monitoring mechanisms and technologies create some alternative possibilities for oversight and control. Specifically:

- Whether through a series of isolated systems or in a hybrid centralised and decentralised system, there is clearly an appetite for household ownership of (some) water to be used for some purposes in some locations.
- A greater challenge may arise if we seek to undo the assumption that human waste should be mixed with water and treated as a liquid. Effective provision for solid human waste would require different collective infrastructure, but may be particularly desirable in dryer locations, and could also (in line with Chadwick's original vision) address some of the fertilisation needs of agricultural crops.
- The denuded state of Bradford Beck highlights the advantages if surface water management is integrated with other processes of water management, as well as showing how local community champions for a watercourse can make a difference.

The third question posed in the introduction to this chapter concerned the processes of change. It asked, insofar as there are different potential ways of managing quality: what do the historic stories about learning to manage quality in a technocratic world tell us about a putative further transition to a more 'reconnected' water management?

The historical story offers some evidence about the role of champions within a transition. Chadwick, Snow and Bazalgette are all lauded today for what are seen as linked contributions to the Victorian revolution in public health. Their role as heroes in popular narratives, however, ignores the substantial and difficult differences in perspective and 'side' between the three men. Chadwick was the visionary: daring to imagine that the state could find a solution to this problem, but obstinate and unreasonable in his insistence on his solution against the evidence and knowledge of those constructing sewers in practice. Snow was also a visionary; however, whereas Chadwick was successful – at least for a period – in selling his vision to the establishment, Snow never achieved such recognition in his lifetime. And this is ironic from our contemporary perspective because Chadwick's persuasive vision was (in many respects) wrong, whereas Snow's ignored understanding was right. Bazalgette was one of the engineers who stood up against Chadwick's impractical universal solution to pipe design, but he also implemented Chadwick's prioritisation of local sewers to the Thames, which we now know perpetuated the city-wide transmission of disease, as well causing the big stink. Bazalgette's true skill as a project manager was only really displayed once the work on the interceptor sewers got underway. Much credit is due to his steerage in enabling this enormous engineering project of London's interceptor sewers to be implemented with public support, even if the justification for doing so continued to ignore Snow's work and repeat the miasma theory of disease transmission. Seen in greater detail, then, rather than three champions riding united towards a sewered city, Chadwick, Snow and Bazalgette pulled in some opposing directions, disagreeing on important details about what should be done, how it should be done and why.

The legacy of this 'messy' transition perhaps suggests that advocates of 'water sensitive urban design', 'sustainable drainage systems' and other innovations in water management should not lose heart. It appears that the real and important differences between perspectives and routes forward form part of the legacy, contributing different aspects of the platform for reform.

Overall, the chapter has shown that, under the technocratic model, water qualities have been universalised into three broad categories of 'potable' piped water, 'clean' surface waters flowing in drains and streams and 'dirty water' containing foul sewage. Even under new, more environmental technocratic management of recent years, these urban waters are still significantly different in quantity and quality to the 'natural' waters that were found before urbanisation occurred. For the ordinary water user, the qualities of different sets of water are relatively unknown and unknowable: trust is placed in authorities, who are usually seen as ensuring that piped water is potable and that if our children paddle in a stream they will not be poisoned. The stories of London's progressive development of sewers show how urbanisation created the pressures for centralisation and control. They also illustrate the messy and political process of change, in which three opposing individuals can only be retrospectively seen as all having moved water management in the same progressive direction. In terms of a direction that might be progressive in the contemporary era, the stories from Waiheke and Bradford illustrate an appetite for greater local knowledge and control, an appetite which would only be enhanced were cheaper and better local monitoring technologies available. If these stories offer a guide to the future, it is for water management incorporating more of a variety of qualities of waters suitable for different purposes, managed at different scales and, hence, for a wider variety of water systems incorporating localities with decentralised systems or elements of decentralisation.

7

WATER OUT OF PLACE

> Flood risk: 'the combination of the *probability* of a flood event and of the *potential adverse consequences* for human health, the environment, cultural heritage and economic activity'
>
> (European Parliament and Council, 2007, emphasis added)

Introduction

Mary Douglas' anthropological studies (1984) famously defined waste as 'matter out of place', highlighting the culturally specific nature of what was and was not waste. Along the same lines, we might define a flood as 'water out of place' – in other words, a flood occurs when water is in a place that we (humans) think it should not be. Compared to the focus of other chapters in this book, the story of how floods are thought about is one in which human perceptions and human practices have loomed large, particularly in the last 30 years. Central to this story is the concept of flood risk management (FRM), the process of managing land, water and people to avoid too many floods causing too much damage. The introduction to this chapter tells the story of how FRM has emerged, four recent case studies of flood risk situations are then used to analyse the nature and state of FRM today.

Daily weather changes, as well as variations between the seasons and from year to year, make 'peaks' in water levels and flows inevitable. Before technology provided the means to manage the water, human lives were managed around such peaks; the oldest roads navigated ridges and hilltops to avoid bogs and stream crossings, old settlements were located in relatively flood-free locations, while herders moved grazing animals between upland and lowland, depending on the season. From the fifteenth century through to the present, new technologies for managing flooding have developed. In cities the problems of a stream or river flooding led to local management strategies such as the construction of flood walls, the deepening and straightening of channels, or – in extreme cases like the city of Palma de Majorca – the construction of whole alternative paths for a river. Coastal communities likewise constructed sea walls and coastal defences or barriers. The philosophy of these 'flood defence' (FD) strategies was to avoid the immediate local impacts of the flood either by moving it away or keeping it away. A parallel approach was used to turn water-logged flood plains and coastal wash-lands into new agricultural land. Unwanted water was removed through drains, sometimes aided by pumps (powered by wind, steam or electricity). These methods were further refined to support the increasing intensification of agriculture after the Second World War (Scarse and Sheate, 2005). Though they required constant maintenance, these techniques were also seen as 'flood defence'.

Alongside FD comes the idea of an FD authority – a body that is responsible for identifying, developing and maintaining FDs associated with a particular source of flooding. Hence, different types of flooding were designated according to the source of the water (see Table 7.1). Different bodies were allocated responsibility for FD,

TABLE 7.1 Types of flood and organisation with oversight for this type of flood

Type of flood/ UK oversight	*Source of water*	*Characteristics*
Surface water flooding *Local authority*	Heavy rain has exceeded the capacity of either the ground or the gulley-pot/drain to absorb it	Can be sudden and localised, hence hard to warn. May only impact a few roads/houses/fields, but recur frequently. Increases with deforestation and urbanisation
Sewer flooding *Sewerage operator*	Heavy rain has gone into the sewer and it overflows into rivers or streets/ homes. May be surface water or foul water sewer	Depending on position in catchment, may be sudden and localised, hence hard to warn. Frequent source of pollution. Often combines with surface water flooding to form 'flash flooding'
River flooding *Environment Agency*	Heavy rain upstream/snow melt means water bursting river banks	Can usually be forecast and warnings given. Tends to have an extensive impact when it occurs, particular in urban areas
Coastal flooding *Local authority*	High tide and/or storm surge overwhelms sea defences and low-lying land floods	Can usually be forecast so warnings given. Tends to have an extensive impact when it occurs
Groundwater flooding in cities *Local authority*	Water table rises and water penetrates localities previously dry e.g. cellars	Long-term cumulative problem, sometimes caused by cessation of water abstraction for industrial purposes
Drainage of agricultural land *Drainage board*	River flood plain or coastal wash-lands protected by sea walls are wet from overflowing rivers	Drains and pumps maintain land so that cultivation is possible

depending on the type of flooding – for example, in the UK, sewer flooding is the responsibility of the sewerage authority while surface water flooding is addressed by the local authority.

In the late twentieth century, critiques of FD approaches emerged. Even from an exclusively flooding perspective, FD strategies were problematic. By building a larger and faster channel in one location, a flood can be hastened downstream, exacerbating the issues for another community. Likewise, the construction of a sea wall and the removal of river or coastal wash-lands increase the flood pressure on neighbouring localities. Additionally, it was noted that these strategies were financially expensive. The development and maintenance of structures countering nature (and nature at its most strong and ferocious extreme) took significant capital investment and maintenance. A further important concern was that drained wash-lands and fast-flowing high-walled channels changed and damaged the environment.

An alternative philosophy for managing floods was developed and is increasingly embraced, at least in rhetorical terms. Rather than speeding the passage of water away and causing flooding elsewhere, an alternative aim is to make appropriate space for the water close to where it first arises. 'Appropriate' because local participation is needed to select from various potentially 'wet' locations to minimise the harm done by flooding. Moreover, once floodable locations are selected, landscape management is then required to ensure that it is these locations that are flooded. The new approach has been given the name flood *risk* management, indicating the aspiration to work with and to manage the risks of flooding rather than to completely prevent them (White, 2010). In this new philosophy, sustainable drainage features alleviate surface water flooding, upland land management combines with sufficient underdeveloped flood plains to relieve river flooding and newly released wash-lands aid coastal flood management strategies. In addition to mitigating the probability of flooding, FRM also includes measures to reduce the impact of those floods through effective preparation. Careful land use planning, combined with warnings, insurance, flood management strategies, planning for exceedance, building and household resilience measures and good emergency services all act to minimise the financial and emotional cost of flooding. 'Making space for water' is coupled with 'being ready for flood' as the new FRM mantras.

The shift in philosophy towards FRM marks a change from just the local management of flooding towards a more integrated approach. Excess water is now to be managed in its totality within the catchment or along the coastline where it is occurring. Human choices about how land is managed are recognised as a critical factor influencing both flood probability and impact. As well as involving the addition of catchment scale considerations, FRM has also brought a shift in the understanding of relevant stakeholders. Rather than floods being just a technical matter of what is likely, and what (technically) can mitigate it, instead FRM is widely recognised now to involve an array of local and community organisations and individuals who have an interest in how land is used, as well as in preparing and protecting those who are potentially vulnerable to flooding. Hence, the scale of consideration is broader in geography (across the catchment) and scope (encompassing many

TABLE 7.2 Measures to address flooding associated with the FD and FRM paradigms

	FD	*FRM*
Construction of hard FDs	Y	Y
Emergency services	Y	Y
Insurance against floods	Y	Y
Land use choices make space for water. These measures include: • upland land use changed to increase storage • judicious location and design of development and redevelopment to minimise flood risk, including use of sustainable drainage • protecting or creating flood plains and coastal wash-lands • planning for exceedance (e.g. roadways are overflow channels when drainage is full)		Y
Flood resilience in locating and designing buildings		Y
Public knowledge of and preparedness for flood		Y

interests, not just flood management), but also more local. (Which areas can be flooded, and which not? How, in a particular location, can flooding be managed?)

As demonstrated in Table 7.2, FRM does not rule out the use of measures formerly associated with FD, such as the construction of hard defences, or the use of insurance to protect against flooding. Rather, it suggests that judicious use of these measures needs to be supplemented with making space for water and ensuring flood resilience. 'Soft', 'non-structural' or 'natural' FRM measures are therefore added to structural measures as the means to address flood threats.

Though the new 'flood risk management' paradigm may make old practices look irrational, it does not offer simple answers about what should happen now. One specific difficulty that it highlights concerns fairness (Johnson *et al.*, 2007; Butler and Pigeon, 2011; Nye *et al.*, 2011). Flooding is inherently unfair: it happens in one place, not in another. But how do we respond? When some people face flooding disaster we may have an intuitively compassionate response: as a government and society, it can be argued, we should share these unfair costs equally through society. The critique of such compassion, however, is that it creates a perverse incentive: we know some locations are more vulnerable to flooding, but unless landowners meet the flood costs they will continue to build inappropriate developments in the wrong locations. Likewise, we know of strategies to minimise the impact of floods on existing buildings and lives, but unless householders meet some costs from floods, (some argue) what incentive do they have to employ such strategies? Added to everyone's busy lives, and a tendency to optimism ('it couldn't happen here'), the governance challenge is to develop a context in which compassion runs alongside shared responsibility, which is accepted and understood, both for minimising flood risk and for meeting flood damage.

Unlike the areas of water supply or water quality, most commentators on flooding recognise and acknowledge the need to see flood in a broader context than just the local control of water that was associated with the previous FD philosophy.

Even where it is widely acknowledged, however, transferring a change in philosophy to practice is problematic. In reviewing four cases of recent flood management, the aim is three-fold. First, the cases each show the constructed nature of flooding, how human processes have layered upon the physical background to create a potential for crises. Second, they demonstrate the opportunities and barriers to the implementation of a new reconnected, people-oriented flood management. Third, the cases provide an opportunity to consider issues around the governance culture through which floods are addressed. In this chapter I focus on describing the cases and considering the processes and outcomes revealed. In Chapter 8, I will compare the same examples with theory about governance cultures and, hence, explore how and in what ways flood governance might change.

It is useful to briefly introduce the process for funding FRM investments in the UK before the case studies are discussed. Management of flooding is centred on the 'lead local flood authority' (LLFA) in an area (depending on the area, this can be either a lower or a second-tier local authority). Insofar as locations are known to be vulnerable to floods, the LLFA can bring forward plans for capital schemes that will address specific local flooding issues. These schemes are subject to evaluation through an analysis that weighs the financial costs against the benefits, the latter perceived principally as the number of homes at risk of flooding, but also taking account of environmental features – for example, whether designated habitats are protected. Approximately an 8:1 benefit–cost ratio is required for schemes to get funded from government (Committee on Climate Change, 2014). A programme of schemes meeting the required ratio is laid out over a six-year period by each of 12 regional flood and coastal committees, with schemes prioritised based on readiness and in relation to local priorities.

River flooding in York

Located on the River Ouse, which drains nearby sub-catchments in northern England (including the Swale, the Ure and the Nidd), York has had a long record of flooding and, hence, provides an interesting example of flood changes through time. Because its main vulnerabilities relate to river flooding, York is a (relatively) institutionally simple area to discuss, and thus an appropriate first case study for consideration. Much of the hydrological material that follows is drawn from Lane (2003), supplemented by the City of York's *Strategic Flood Risk Assessment* (City of York, 2013) and personal communication with flood risk managers in York.

In November 2000, the water levels in floods in York were among the highest ever recorded, with the river 5.4 metres above its usual level, and the river discharge calculated to be 11 times its usual flow. The then Deputy Prime Minister John Prescott took these floods to be confirmation of a change to more frequent and higher-magnitude events that occurred as part of climate change. The work of Lane (2003) enables us to unpick this claim, and also to understand something of the flood history of York.

York has a record of flooding dating back to 1263, with recent severe flood events occurring in 1947, 1978, 1982, 2000, 2012 and 2015. In the distant past, flooding involved both the effects of rivers and those of tides; but with the construction of a downstream weir on the River Ouse in 1757, the tidal influence was effectively removed. Data on flooding in York is among the most enduring in the UK, with a consistent record from in 1878, hence strengthening the reliability of any conclusions about long-term flood frequency and magnitude.

Understanding changes in flooding is usually achieved by looking at changes in the number of flood peaks that exceeded a particular magnitude over a set period. This is often supplemented by information on the maximum flood level for each year. Results from Lane (ibid.) show a dramatic rise in the number and magnitude of floods in York during the 1940s and again in the 1980s and 1990s. Questions are then raised over the cause of these changes.

Lane highlights the range of different factors that could influence the flood record. One possibility is that floods have become greater due to changes in precipitation. For example, there might be more precipitation, or it might arrive more intensely, or it might be sequenced differently in terms of the different tributary catchments. Another set of possibilities relate to the processes of water retention or discharge within the catchment.

Through statistical investigation of these different possibilities, Lane concludes that the central influences upon changes in the record of flood frequency and magnitude in York appear to relate to the changes in land use. Specifically, he identifies the increase in upland drainage during the 1940s as a key influence upon the increase in flood frequency and magnitude. Draining peaty upland soils makes the land more usable for stocking purposes, but it also decreases the extent to which water is retained in the catchment, leading to increases in flood peaks. In relation to the changes in the 1980s and 1990s, these are understood to be linked to altered stocking practices, as EU subsidies drove increases in the numbers of sheep in upland locations. By reducing biomass and compacting soil, increased stocking is known to reduce the extent of upland infiltration, resulting in more overland flow and, again, an increase in the sharpness of flood peaks.

As Lane (ibid.) notes, by the time he wrote, some of the problematic changes to upland management were beginning to be reversed. Experiments were being carried out on the removal of drains to aid upland retention of water. Also, a combination of the foot and mouth epidemic in 2001 and changes to the farm subsidy regime meant that the ability and incentive to over-stock uplands was much reduced. In the decade since Lane wrote, the latter incentives have continued. Government policy acknowledges important ecosystem services – including both water and carbon storage – performed by the upland, and hence aims to focus subsidies to support and retain these functions (Defra, 2011). The Environment Agency's *River Ouse Catchment Flood Management Plan* (EA, 2010: 13) reinforces these changes, stating their intention to 'work with landowners... to change the way that land is managed and slow the rate at which floods are generated' and also to 'investigate creating flood storage to manage flood risk'.

Although these developments are welcomed by York flood managers, their precise pace and impact remains uncertain. Moreover, as Lane acknowledges, while upland land management may help address the extent of flooding problems in York, climate change means that the city must strengthen its own resilience in light of the high probability of higher future discharges. According to the City of York's *Strategic Flood Risk Assessment* (2013), Defra suggests that climate change means that peak flows in rivers should be expected to potentially increase by up to 20 per cent over the next 50 years. Old defences, formed by earth embankments, brick or stone-clad concrete walls and gates, are not believed to meet the required one in a 100-year protection. Moreover, in addition to those areas directly vulnerable to flooding from the Ouse, areas upstream from the Ouse FDs also have the potential to flood from the build-up of surface water, river, sewer or foul sewer flows behind defences. Each defence is therefore associated with a pumping station that pumps flood build-ups behind defences into the swollen main river during periods of flood.

As a consequence of the significant flood risk, all proposed developments in York are subject to careful consideration in relation to both the development's own vulnerability to flooding and its impact on others' flood risk. The Council's *Strategic Flood Risk Assessment* classifies areas according to their level of risk and lists policies to be followed accordingly (City of York, 2013).

Surface water has not been the cause of significant problems in York the past. However, following severe surface water flooding in Hull in 2007, the government insisted that local authorities reviewed their area's vulnerability to surface water flooding through production of a surface water management plan. York's plan (City of York, 2012: 2) highlighted how the City of York lacks knowledge of the surface water management assets, and also states that the maintenance of existing assets is poor, arguing that 'repairs to drainage systems ... have often been ill thought out with no regard to a holistic solution'. The study also showed how development, and particularly highways and highway drainage, can have a significant impact on surface water flooding within a flat area like York. A commitment has been made by the Council to develop further investigations of the surface water drainage assets and to incorporate a better understanding of water management and drainage planning into the processes of highways planning and development.

So, to what extent can we conclude that FRM in York follows the full holistic, environmentally and socially integrated approach suggested by FRM? The measures utilised in York are summarised in Table 7.3. Overall, the York case seems to show some success in working towards joined-up policy and institutional thinking and acting between agencies. For example, the EA's catchment management plan systematically encourages upstream management of water in order to relieve downstream flooding. Likewise, the City of York Council policies feature very careful checks to ensure that new development does not increase the flood risk. It also should be acknowledged that such joined-up institutional thinking incorporates significant agency-specific interactions with the public. For example, Natural England will interact with farmers about the nature of their upland subsidies, York planners will interact with people making developments, while the Environment

TABLE 7.3 Summary of FRM measures utilised in York

	FD	*FRM*	*York*
Construction of hard FDs	Y	Y	Yes – hard defences are in place and extended
Emergency services	Y	Y	Of course
Insurance against floods	Y	Y	Of course
Land use choices make space for water		Y	Yes, particularly in relation to uplands, but also where possible close to the city
Flood resilience in locating and designing buildings		Y	Yes, in relation to building location No, in relation to resilience in existing buildings

Agency will provide flood warnings to those York households vulnerable to river flooding. However, while the surface water management assets are acknowledged to be little known and poorly maintained, it is not clear how these factors will be made good without significant new resources. In this respect, joined-up activity cannot make up for lack of resources. It is also notable that York City Council's website does not mention flood preparedness or flood resilience. Hence, while various sets of the public are involved with flood prevention in York in their role as farmers, or developers, the general residents of York – particularly those vulnerable to flooding – are not apparently being actively engaged in the process of flood management.

Surface water inundation in West Garforth

West Garforth is a neighbourhood of a few square kilometres in the east part of the city of Leeds, in the North of England. Parts of the area have been subject to surface water flooding since the 1980s, which residents perceived to be getting worse. By the mid-2000s, some properties were vulnerable to flooding as frequently as every two years; in this respect, we can see the flooding problems as chronic. These problems were not all adjacent, but dotted through the area. Despite various attempts to address the problems from the 1980s through to the 2000s, solutions to the issue of flooding in West Garforth proved difficult to develop because they were expensive and involved overlapping jurisdictional responsibilities between householders, the local authority and the water company. For this reason the area became the focus of an 'integrated urban drainage pilot project' sponsored by Defra, involving academics, the Environment Agency and Leeds City Council, and running between 2006 and 2008. This section uses the story of the pilot project to draw conclusions about neighbourhood-level FRM. The information below is drawn from the Defra report (Defra, 2008), supplemented by personal knowledge.

The integrated urban drainage pilot projects asked 'What can be learnt through integrated working between agencies?' Hence, the West Garforth pilot sought new ideas about how the problem could be addressed, though it did not have the capacity to actually implement solutions to the flooding identified. All the pilots contributed learning to the new FRM regime in the UK that was enacted in 2010 through the Flood and Water Management Act. Subsequent actions to address flooding problems in West Garforth shed light on the extent to which this regime is achieving success in relation to its chronic surface water flooding problems. For brevity, the story of the West Garforth pilot recounted here focuses on the northern part of the study area (ibid.). The case helps to demonstrate the processes through which surface water flooding problems can arise, the institutional challenges of tackling these neighbourhood-scale issues and the extent to which the UK's new FRM regime has enabled surface water management issues to be addressed. The conclusion to the case also reflects on the extent to which the issues described are paralleled in other jurisdictions beyond the UK.

A map of West Garforth from 1850 shows a rural area intersected by one main watercourse with a number of tributaries. During the subsequent 150 years, the suburb developed in an ad hoc way, with one-off house or street-level building projects giving way to infill developments in gardens, or, more recently, the replacement of large properties with more dense housing. The proliferation of impermeable roofs, roads and driveways means that rainfall flows quickly over land towards watercourses or sewers rather than slower processes that occurred formerly, of water infiltrating and flowing through the soil. Positioned between the backyards of different properties, the original watercourse has been covered, but continues to form the backbone of the drainage system. This has become problematic as there is no manhole access, and, because of the position of the houses, there is little chance of achieving machine access to create a manhole in order to inspect its quality or to address blockages. The culverted nature of the watercourse also means that building owners may be unaware of its existence; in any case, they lack the access they would need to enable them to discharge their responsibilities as 'riparian owners' to keep the watercourse free from blockages.

The project team's difficulties collating data illustrates the problem of understanding urban issues when surrounding responsibilities are fragmented. Even to characterise the existing drainage system involved combining geographical maps in different formats belonging to different organisations, and using CCTV surveys of the gullies and 'lidar' studies of the road levels to supplement and make good gaps in this data. Information about previous flood incidents was also challenging to collate; different bodies had different information, while organisations' fears about compliance with consumer protection legislation delayed the process of sharing. In addition to the collated information, previously flooded members of the public were invited to a public meeting to explain their experiences of flooding in relation to six flooding locations in the context of detailed maps. The information from this first public meeting further validated and developed the team's understandings of past flooding.

Modelling current and future flooding involved the standard engineering process of constructing a computer model to calculate flows on the roads and through the underground drainage system in the context of different 'design storm' events (a design storm is a standardised algorithm which mimics a real storm which is likely to cause flooding). The accuracy of the models was checked through comparison with data about how the system had behaved in past storms. In addition, serious flooding in West Garforth took place in summer 2007, in the middle of the project study period. Because of the preceding public meeting, many members of the public recorded detailed information about the direction and path along which the water flowed during this event. Volunteers from an action group formed at the meeting also conducted house-to-house surveys that revealed more properties subject to internal flooding than the official records have shown. This information was invaluable in increasing the accuracy of the model.

Exploring potential solutions to flooding drew on the understanding of both the project team and the public. Potential interventions involved, variously, the development of surface storage, the addition or enlargement of pipes or the disconnection of certain areas from the drainage system. At the second public meeting, residents were asked to select their preferred locations for new drainage features on large maps of the area, drawing on their knowledge of how the area was used. Their ideas were then combined with those of the project team to be modelled and costed, resulting in a set of interventions ranked in terms of their costs and benefits at present and in the future. The measures recommended to give the most benefit for their cost included the use of a diversion pipe to reroute some highway drainage into a different drainage basin, the potential for re-engineering a playing field to act as water storage when rainfall had been heavy and the possibility of using some wide upstream streets as localities for street-scale sustainable drainage (i.e. local storage). While no combination of solutions was expected to completely prevent future flooding in the area, the study showed that the expected future damage could be significantly reduced. Though direct funding for these recommendations remained elusive at the end of the project, they formed a 'short-list' of measures to be further explored and developed.

The investigation also led some 'easy wins' to be identified. The CCTV survey showed where other services had dissected gullies, blocking the path of water, which could then be addressed by the Council. The information provided by the study was also crucial in enabling some modest investment decisions by the Council highways department and the water company to address flooding at specific points in the network. An additional, unexpected positive outcome of the process was the formation of a local flood action group of residents who had been subject to flooding and were keen to collate information and campaign for flood management action to occur. This group was the direct result of the public meetings in the project, and continues to operate more than six years after the project was completed.

The project made a series of recommendations about the process of surface water management in the UK that informed the development of guidance on surface water management plans and the content of the 2010 Flood and Water Management Act, as well as a significant shift in the flood management regime in the UK. An important change since 2008 that will prevent the further deterioration of surface water drainage all over England is that the public are no longer permitted to cover gardens or driveways with impermeable surfaces without planning permission. Moreover, all new developments are required to demonstrate how they will impact on the drainage system, and to compensate with site-level storage if the system is close to capacity. These changes mean that future demand for drainage capacity has – appropriately – become regulated as part of the planning system. In England, at least, we can conclude that there will be no 'new' West Garforths.

Nevertheless, addressing historic problems like those in West Garforth remains difficult. Changes since 2008 permit local authorities to require cooperation from water companies and the Environment Agency in the development of surface water management plans. Effectively, whereas previously it was beyond the scope of their responsibilities, local authorities are now empowered to lead studies like that undertaken in West Garforth. A further area of change is that authorities are now required to keep a list of drainage assets in their area, including ownership and state of repair. This register is intended to inform a structured and risk-based maintenance schedule. However, a significant issue relating to both surface water management plans and the register of assets is the lack of resources available to local authorities. Severe cuts since the legislation was passed mean that, although local authorities are *empowered* to do more investigation and maintenance, undertaking such work has to compete with other core duties such as education or social services. In practice, until areas experience severe flooding, it is unlikely that local authorities will have invested resources to have a full register of assets.

In terms of funding for specific drainage schemes, the potential for supporting drainage work in West Garforth has increased since the UK government introduced the 'partnership funding' scheme in 2012. Before 2012, schemes either achieved a high enough benefit–cost ratio to qualify for government flood grant in aid, or they received no funding at all. Under the new arrangements, a range of local contributions can lower the proportion of the costs needed from government – for example, partnership contributions might be secured from the local authority, from the 'local levy' (money paid by local authorities into a regional fund and distributed by a regional committee), from the water company, or from business/householders in the area.

At the time of writing, two new schemes are helping to address flooding in West Garforth. The first small scheme sought to increase the extent of storage possible on the local football pitch, reducing the flashiness of downstream flow. Funding for the £80,000 scheme was obtained through a combination of local authority sports-facility related investment and money from the local levy. In the second scheme a hard engineering diversion from the existing culvert has been created at a point in the culvert that was particularly lacking in capacity and denuded; this investment cost £104 million and was funded through a combination of (national) 'flood-grant-in-aid' and (regional) local levy. While these investments are likely to address some of the chronic problems of West Garforth, they do not represent a complete solution to the problem, particularly given climate change-related expectations about the changes in the rainfall intensity of extreme events.

Finally, there are also a number of difficulties associated with the process of addressing surface water flooding, illustrated by West Garforth, that have not been overcome through legislative changes. Notably, the lists of assets and their ownership does not overcome the difficulty of the group of 'riparian owners' of the culvert in West Garforth and many other similar drainage assets, who may be unaware of the asset,

TABLE 7.4 Summary of FRM measures utilised in West Garforth

	FD	*FRM*	*West Garforth*
Construction of hard FDs	Y	Y	Yes – new culvert can be seen as a hard defence
Emergency services	Y	Y	Of course
Insurance against floods	Y	Y	Of course – though not available in highly flooded locations
Land use choices make space for water		Y	Yes – for example, in relation to football field
Planning for exceedance		Y	Yes, in relation to football field
Flood resilience in locating and designing buildings		Y	No, in relation to building location Yes, in relation to resilience in existing buildings, which is supported through the local flood action group

and/or incapable physically or financially of discharging their supposed responsibilities. Moreover, the problem of under-reporting of surface flood events that was demonstrated in the West Garforth case is likely to remain, with websites still advising the public to select whom they call according to the source of the flooding. There has also been no progress in addressing householders' potential concerns that reporting flooding could impact on their home's value and its treatment by insurance companies.

The West Garforth case, the characteristics of which are summarised in Table 7.4, can be seen as an illustration of causation of surface water flooding problems and the consequent severe difficulties of addressing these problems in terms of retrofitting drainage capacity. For countries that are currently urbanising, it provides a stark reminder of the need for the drainage processes of urbanisation and infill development to be part of a planning system in which surface water drainage is given explicit consideration.

The pilot project in West Garforth can be seen as something of a foretaste of the surface water management regime that has been introduced across Britain. Can we conclude that this regime 'reconnects people with water'? Surface water management issues are certainly better addressed than they were in the past, with local authorities named as responsible authorities, which includes the discretion to work with local people if they wish. Of course, like investment in research about a flooding problem or the design of a scheme, such costs undertaken to consult the public ahead of a scheme have the potential to lower the resources available for construction, but this may be a small price if the scheme becomes more effective at avoiding flooding, and/or if it better fits other local needs. Importantly, however, funding for such schemes remains to a large extent in the gift of central government directed through an algorithm based on value of properties at risk. A particularly interesting consequence in West Garforth was the formation of the local flooding group. While the investigators in the pilot project and the local authority team tended to be focused on interventions that would 'solve' the problems, the local flood group have developed other activities that support local resilience, such as providing advice to householders about minimising flood damage, and local flood warning systems so that they can all be alert when potentially threatening rainfall is forecast. Hence, insofar as the new regime allows surface water management to occur, and if the local authorities use their discretion to work with local people, it appears that, yes, the new regime does have the potential to be inclusive and discursive about interventions.

Rural flooding in the Somerset Levels and Moors

> Few places better illustrate the farming, flooding, conservation dilemmas than the Somerset Levels and Moors.
>
> (Hicklin, 2004: 20)

> 'We cannot let this happen again.' UK Prime Minister David Cameron commenting on the flooding of the Somerset Levels in February 2014.
>
> (Smith *et al.*, 2016)

From December 2013 through to the end of February 2014, flooding on the Somerset Levels and Moors drew the attention of the UK national press. The story was not just the human and animal suffering, but rather the recrimination that residents and some government ministers levelled at the Environment Agency. This played out a stereotype of urban versus rural interests: the conservation and urban-orientation of contemporary flood and water management was caught on the back foot, critiqued by a 'rural' perspective that stresses the importance of farmland, food production and the cultural landscape of agriculture. It provides a useful case to consider here, not only because its rural nature contrasts with the urban examples discussed above, but also because, as a low-lying coastal area, it brings in the additional consideration of the sea and the prospect of sea-level rise.

The Somerset Levels and Moors are some 650km^2 of low-lying coastal land on the West of England (Acreman *et al.*, 2011). At or slightly above mean sea level, the area can be subject to tidal flooding as well as to inundation from winter floods. The area's wet origin is indicated in the name 'Somerset', meaning summer people, and demonstrating that, as a peat-forming fresh water marsh, this area was occupied by prehistoric people only during the dryer summer months. Its marsh status was progressively changed through drainage, from medieval times until the nineteenth century (ibid.), and then further 'improved' by lowering some water levels (and consequently ground levels as the land dried out) through pumping in the 1970s and 1980s. River channels are canalised to maintain drainage, and water is raised into these above-ground channels in order to ensure the water flows out to the sea. Tide gates or barrages act as a barrier so that high tides cannot easily flow up the rivers and flood the area. However, in recognition that the new management regime was causing the loss of valuable wildlife habitat, since the 1990s there has been action to reverse some of the changes brought to the area during the 1970s and 1980s.

TABLE 7.5 Potential management regimes for the Somerset Levels and Moors

Regime	*Management*	*Implications for*		
		Farming	*Nature*	*Flood*
Farm	Winter – pumping from fields to rivers Summer – river irrigates fields. Rivers dredged to maximise flow	Yields maximised, but some loss of peat over time	Winter – lack of standing water for migrating birds Summer – lack of dry meadows for summer flowers	Farm flooding minimised hence more flood for rivers/urban areas
Nature	Farming is abandoned and the area is allowed to revert to a fresh water marsh	No farming	Revert to biodiversity rich fresh water marsh	Marsh acts as a sponge for flood water
Culture	Mid-twentieth-century management regime is maintained with limited pumping and dredging	Much lower yields than now	Avoid damage to nature wrought by 'Farm'	Occasional inundation is inevitable to minimise flooding elsewhere

Source: Author's table, with information drawn from Hicklin, 2004, and Acreman *et al.*, 2011

Commentators on the region make it clear that there are choices over different management regimes that have the potential to yield different benefits (Hicklin, 2004; Acreman *et al.*, 2011). These are characterised in Table 7.5.

In Table 7.5, the 'Farm' regime represents the extreme of a food production-oriented means of managing the land and is similar to that practised in the 1970s and 1980s. At the opposite extreme, to follow the 'Nature' regime would be to abandon most of the active management and to let the land resume its 'natural' functions, supporting diverse wildlife, as well as acting as a sponge for fresh water or tidal flooding. The third, 'Culture' regime would seek to preserve the water management techniques of the early twentieth century; this would maintain the internationally important wildlife habitats (Hicklin, 2004), while keeping the cultural landscape of farming. Either 'Nature' or 'Culture' could form part of an 'FRM approach' as described at the beginning of this chapter.

Following this framework, the period from the late 1990s to the present can be characterised as 'Farm' giving way to 'Culture'. However, there was no smooth clinical transition. As Hicklin (ibid.) (the local Environment Agency customer manager) acknowledges, the subsidies for environmentally sensitive farming through the 1990s were not well designed for Somerset farmers. Specifically, they required land to be contiguous and within one hydrological unit, which was not possible for most farmers to comply with; on the flat land of the moors and levels even very small changes in height lead to different drainage processes, hence one block of land can form separate 'hydrological units'. Consequently, until the early 2000s, there was little subsidy to sweeten the costs to farmers of giving up on the previous lucrative farm regime. And even through his talk of a 'new consensus' around the management of water, it is possible to discern Hicklin's frustration, tinged with understanding about what he calls the ingrained 'flood and drainage cultures' of the local Drainage Boards, the Environment Agency operational staff and the farmers. For example, he suggests that the latter are disinclined to 'farm water' (and subject themselves to prolonged inundation) because 'they fear that long-term viability of their land will be irretrievably compromised by soil and vegetation-structure changes' (ibid.: 24). Hence, while nature conservation-minded theorists like Acreman and colleagues may paint a clear picture of trade-off between the 'ecosystem services' of nature conservation and (external) flood alleviation, on the one hand, and agricultural production, on the other, it is clear that transforming this theoretical trade-off into economic instruments that are effective and trusted by farmers has not been easy.

These tensions between the 'Farm' and a more flood and nature-friendly 'Culture' management regime can be seen as the backdrop to the flooding of 2013–14. According to the *Somerset Levels and Moors Flood Action Plan*, from March 2014 (Anon, 2014: 1),

> The prolonged wet weather and subsequent flooding in Somerset began in mid-December 2013. Within the Levels and Moors over 150 properties are now flooded internally and 11,000 hectares of agricultural land remain under water. Over 200 homes in several communities have been cut off, some for more than two months.

The damage wrought was, therefore, that some rural properties were internally flooded and others had their access cut off for a prolonged period; additionally, a large amount of farmland was inundated for more than three weeks, causing damage to vegetation and soils which can take up to two years to recover. The difficulties prompted criticism of the ['Culture'] water management regime that was accused of prioritising conservation over flood prevention, and favouring protection of urban areas over rural districts.

The *Flood Action Plan*, quoted above, was the product of a rapid local participatory discussion. Actions proposed to remediate and prevent the repetition of the event include new processes for the management of rivers, and extra resources to support catchment-based flood planning. Hence, for example, dredging was reintroduced for the first time since 1998, meanwhile the catchment-sensitive farming regime that had previously focused only on improving water quality would be piloted for application to flood prevention as well. The potential for concerted action in uplands to impact on lowland flooding has been demonstrated by research by Lane and colleagues in Yorkshire (Lane *et al.*, 2011). Such actions were only possible because of the exceptional circumstances wrought by the flood: at least in March 2014 David Cameron's claim that 'money is no object' offered the potential for new resources to support the Somerset Levels and Moors. Despite the polarised media attention, we can see that the routes forward tread a path between the 'Farm' and 'Culture' routes: while dredging is undoubtedly a victory for a 'Farm' management regime, more resources for catchment-sensitive farming is rather a question of resourcing and supporting the 'Culture' approach.

The prolonged media attention to the Somerset Levels and Moors can partly be seen as a function of national politics. Rural areas like Somerset are an important constituency for the Conservative Party, the major player in the Coalition government in power from 2010–15. Nevertheless, the institutional architecture of FRM and nature conservation is the legacy of the previous [arguably more 'urban' and more 'green'] Labour government. And yet, apart from reducing funding, the Coalition government had done little to change this regime and its reduced emphasis on draining rural lands. In this context, the Somerset Levels flood offered an opportunity for rural interests within the Conservative Party to assert their needs and to challenge the Coalition's acquiescence to the approach to flood management ('Culture') focused on balancing farming and nature conservation. The extent of flooding in Somerset also needs to be put in context alongside other flooding incidents. By way of comparison, the 150 properties flooded in Somerset contrasts with the 8,600 homes flooded in the city of Hull alone in June 2007 (Whittle *et al.*, 2010 – see next subsection).

The Somerset case provides a demonstration of the challenges of moving towards a FRM regime. If one considers the number of homes damaged or the amount of economic disturbance created, a greater emphasis on flooding rural not urban areas is appropriate and sensible. However, the change towards FRM involved a shift away from the FD subsidy to farmers that was effectively part of the late twentieth-century flood management regime. It is hardly surprising that this change is resisted. It is also clear that the development of a new flood management approach is still a work in progress; while the principle of paying for the ecosystem services delivered by the farmers through their management of the land is well understood, making such transactions real is difficult and faces many problems. As Hicklin highlighted, rules about the size of plots and their hydrological integrity effectively excluded many farmers from securing environmental subsidies from the late 1990s. More recently, the catchment-sensitive farming schemes have concentrated on water

TABLE 7.6 Summary of FRM measures utilised in Somerset

	FD	*FRM*	*West Garforth*
Construction of hard FDs	Y	Y	Yes – dredging
Emergency services	Y	Y	Of course
Insurance against floods	Y	Y	Unlikely to be available in future
Land use choices make space for water		Y	Yes, in terms of seeking to retain more water in the uplands
Planning for exceedance		Y	Yes, exceedance leads to flooding of the countryside, as in 2014–15 floods
Flood resilience in locating and designing buildings		Y	Historic building pattern puts buildings in more resilient locations. More building unlikely to be permitted

quality, but have not (until the pilot following the Somerset flooding incident) focused on the management of the quantity of water in the system. While, no doubt, schemes can be designed more effectively than they have been, their development requires a complex and difficult compromise that is unlikely to ever be seen as satisfactory by all parties. The Somerset floods show that those who lose in any readjustment of flood policy are likely to be unhappy, and to complain loudly when they are visited by the flood consequences of that policy, particularly if it coincides with a moment when they have a political advantage.

Flood recovery in Hull

Here, the story is not of the prevention of flooding, but rather that of flood recovery following the flooding of 8,600 properties in Hull in June 2007. Drawing on the research carried out by Whittle and colleagues (2010), the account helps us to understand the experience of flooding and the crucial aspect that flood recovery has to play if we are to have a system of flood management that is fair and 'cares'. The research involved focus groups and documentary analysis, with 40 diarists among those who had been flooded in Hull alongside those who were front-line workers dealing with the effects of the flood. The research provides what is sometimes a harrowing account of how people – some of whose lives were built on a precarious balance of local support – were thrown upside down through the displacement caused by flooding. Here, only a taste of what was found can be presented. The question it raises for the wider account is whether and how FRM supports – or could constitute – a more caring flood recovery process.

Hull, like the Somerset Levels, is partly below sea level. It relies on pumps to lift water in the low-lying areas up to rivers so that it can flow to the sea. The flooding in 2007 occurred because the large amount of rain in the vicinity overwhelmed the drainage system and the pumps such that water accumulated across the area (Pitt, 2008). The pattern of flooding was complex, therefore, with many different 'pockets' of flooding where water accumulated when it could not be lifted out of the system. The flooding was experienced in a wide variety of ways, including water rising and seeping in through doors and windows, water rising through the floor, or toilets overflowing and unable to flush.

Interestingly, one of the difficulties encountered by the participants in the flood recovery research concerned the definition of a flood. Where water entered the interior of people's houses there was no doubt – but other cases where damp courses were breached or, somehow, water entered the fabric of the house unseen – and damage emerged slowly, over a period of weeks or months after the flood – were far more uncertain. Owners of these houses had, at first, thought of themselves as lucky compared to their flooded neighbours, but subsequently found that help and insurance claims were refused because the evidence that they were flooded was more ambiguous.

The research highlighted how the process of flood recovery involves significant work for residents in contacting and managing loss adjusters, builders, councils and other services, as well as replacing lost goods. This work has to be done alongside normal life (school, work etc.), often while living in unsatisfactory temporary accommodation, and with networks of support disrupted by flooding (e.g. family who do babysitting are no longer local). One central message arising from the research is that while being flooded is horrible, recovering from being flooded is even worse. Particularly difficult times include watching possessions and an old and loved house be 'stripped out', negotiating with bureaucratic organisations for your basic needs to be met, being physically displaced during repairs, moving back into a recovered home and finding things different or improperly fixed, and coping with the fear of future flooding whenever rain occurs. Many people experienced stress, family issues and mental health problems as a result of these difficulties well over a year after the flood 'event'.

The research suggests that 'flood vulnerability' is about both vulnerability to the fact of being flooded and what they call 'vulnerability to its impact'. This concerns the coping mechanisms and breadth and depth of social and financial resources available to fight for resources and to achieve rapid recovery. In terms of vulnerability to flooding's impact, groups that suffered particularly were the elderly, those with young children (particularly single parents), those with a disability and those living in private rented accommodation. It was noted, however, that these were easy categorisations, and that the extent of people's coping mechanisms related to a range of other factors such as their expertise (for example, in building), or their financial and social resources. It was also a dynamic thing, related to other events and processes in people's lives – so vulnerability is magnified if there are family crises, mental health issues, or illnesses to address during the recovery process. In this respect, the research argued that the front-line workers needed to be given the flexibility to define vulnerability as they saw fit, rather than relating to fixed categories that did not recognise people's special circumstances.

Disasters like floods are often assumed to help bring communities together. The experience in Hull suggests that, though people experienced community cohesion during and just after a flood, this is not always maintained. Particular tensions appear between different groups – for example, the flooded and the non-flooded, the insured and the non-insured, the self-menders and those at the whim of councils or landlords. The stress of being flooded goes on and on, but, after the initial storytelling, people who have been flooded feel guilty about telling their friends and relatives that they are still struggling months after the flood, so they repress the thought, further increasing their isolation. Some flooded people found themselves the object of jealousy because of the perceived benefit of insurance payments and a newly redecorated home. The research suggests that such jealously was almost entirely misplaced, while for a few diarists the replacement and upgrading of their home did work – partly – to their advantage, for most people the personal and financial costs of the flood were huge, and far greater than any benefits.

Interestingly, it appears that the process of the research carried out useful work for policy-makers and the society of Hull in terms of enabling feedback and giving some of those who were cut off from support a voice. Two things stand out.

First, publicity about the research brought a series of letters to the research team, highlighting further or different aspects of flood stress that they had not previously been aware of. (For example, a letter told the story of a family who though not flooded themselves, had to accommodate the flooded mother-in-law, who caused great disruption to family life.) Second, a workshop after a year of the research brought some of the diarists together with those managing and organising the flood recovery. Through careful timing and facilitation, the event became a forum for learning about how floods could be better managed in the future. This shows how the very existence of the research was seen to give a voice to those who were suffering, and enabled further hidden voices to be heard.

Though the Council and the government took moves to promote flood-resilient rebuilding, this was not effective in Hull. One issue concerns tradespeople who were not familiar with flood-resilient building techniques and, in the context of great pressure on their work, were not inclined to try something new and different. Another significant barrier to achieving a more flood-resilient building stock was the insurance industry, however. The insurance industry would not pay the extra costs incurred to ensure flood resilience in a building, nor would they lower insurance premiums if a building had been fitted to be more resilient.

While previous subsections in this chapter ended with a review of flood measures utilised and their fit with FD and FRM, such an overview is not appropriate here, where the focus on flood recovery goes beyond the temporal scope of either process. Perhaps the most telling lesson from this research is about the need for services of flood management and flood recovery to be organised with an ethic of care. Of course services are over-stretched at the time of a large flood, but flexibility and willingness to listen and adjust processes and procedures can go a long way to helping people feel like they have a stake in society.

Conclusion: FRM – a work in progress?

At the beginning of this chapter it was suggested that FRM is the area of water management in which 'reconnecting people with water' was most developed, compared to other areas of water management. In rhetorical terms, at least, it was argued that academics and FRM authorities were widely promoting an idea of FRM that could be distinguished from FD in seeking to 'make space for water' and also 'prepare for flood'. Having looked at how flood risk is being managed in four locations across the UK, this conclusion aims, first, to explore what the cases tell us about the implementation of FRM in the UK and, second, to take brief stock of actions on FRM in other locations as a benchmark to position the UK's actions in a global context.

The four cases discussed have all emphasised how human development processes have unintentionally exacerbated the problems arising from the inevitable physical variation in rainfall, stream flows and tides. Some of the human development processes that have contributed to flooding problems in each of the cases discussed are shown in Table 7.7. One significant change in understandings associated with the

TABLE 7.7 Human factors recognised to contribute to flood risk

Case	*Human factors recognised as contributing to flood problems*
York	• Changed upland land management • Construction in the flood plain • Lack of coordination between agencies
West Garforth	• Increasingly dense urbanisation, including concreting of driveways • Problems caused by systems working together, but limited (previous) opportunity to study how they interacted • Legacy culvert without access arrangements, and where legal responsibilities are allocated to riparian owners who may lack knowledge of their responsibilities and certainly lack the capability to address any problems
Somerset Levels and Moors	• Different views about appropriate management regimes for low-lying wash-lands – specifically, the extent of pumping and dredging • Upland management practices
Hull – flood recovery	• Multiple agencies juggling or avoiding responsibilities for flood recovery processes • Fixed definitions (e.g. of a vulnerable person) with insufficient flexibility for front-line workers to designate people according to perceived need • Lack of insurance support for resilience measures

shift from FD to FRM is that these factors are now recognised as a cause of flooding and a legitimate object for management strategies. But herein lie two challenges. First, flooding crosses geographical and functional boundaries. Whereas FD involved neatly demarcated understandings of responsibilities, FRM's broader understanding of flooding brings with it a need for coordination. How can different areas of policy handled by different arms of government or the private sector be coordinated to deliver an effective collective service? Second, the influences on flooding, as well as its impacts, relate not just to the behaviours of government agencies, but also to the actions of wider publics. How and to what extent are these being marshalled and managed? The discussion below will deal with these challenges in turn.

The cases discussed in this chapter demonstrate some significant successes in relation to coordination. In York, interaction with Natural England has pioneered changes in upland farm management practices and, likewise, the integrated study carried out in West Garforth provided a model that informed the development of the new Flood and Water Management Act through local authorities working with water companies and academics. Even in Somerset, the legacy of the 2014 floods was some clear coordination between parties, using mechanisms designed to address water quality to also address issues relating to water quantity. But we can see that much of this coordination is post hoc – it only occurs after the realisation that there is a problem. There are also ways in which even in these areas, coordination is not happening.

One specific policy area of concern highlighted in the York, West Garforth and Hull case studies relates to urban surface water management. This is partly a legacy issue – little attention has been paid to the local culverts, drains and sewers through which water is collected from urban areas for two centuries of development. But these poor-quality and unseen features remain the responsibility of multiple different owners who often lack knowledge about, motivation or the means to address problems. While there are now measures in the UK for local authorities to obtain knowledge and to carry out action when acute problems occur, there is little opportunity to address smaller chronic problems. Moreover, severe austerity-driven cuts in local authorities mean that monies allocated by government to support FRM roles can easily be absorbed in other areas of policy action (Committee on Climate Change, 2014).

More broadly, though the cases exhibited reasonable coordination between flood authorities, evidence of coordination beyond the flood agenda was rare. In the Hull case study, coordination was demonstrated with social services during and after the acute flooding events that occurred and, similarly, scientific and research agendas were part of the process of addressing flood risk in the West Garforth and Hull cases. But it is clear that this area of cross-agenda coordination is challenging. Notable important organisations that impact on flood risk planning and management are the local planning and highways authorities, the local social services, other biodiversity services and local wildlife groups. It seems that while, in theory, there is enthusiasm about developing spaces so that they achieve multiple functions, most funding regimes are not oriented to support such processes. Hence, for example, a wildlife-rich play area that also achieves flood storage would require that flood management monies were combined with other funding associated with biodiversity, and grants for local play facilities. Agreement would have to be found so that each funding body could be satisfied that their 'client' needs were met, while management was achieved so that all three functions would be sustained over time.

When it comes to coordination with other agendas, a contrast can be made between the UK's approach to FRM with that in the Netherlands. Here, the significant national programme called 'Room for the River' prioritised the large-scale implementation of natural FRM measures. Stimulated by near-miss flooding events at home and the devastation of Katrina overseas, a set of 30 related large projects sought to mitigate flood risk, while also addressing local issues, such as maximising biodiversity or maintaining heritage landscapes. The programme represents what is generally seen as the successful reconciliation of flood management, nature conservation and local planning agendas (Wesselink, 2007; Van de Brugge and Rotmans, 2007). (See, however, the discussion of flood resilience in the Netherlands below.)

Moving on to the second challenge above, the public have various potential roles to play in relation to FRM, but all require active interaction in ways that did not occur under FD: they are stakeholders in land that can help prevent flooding, they are also owners or workers in properties potentially subject to flooding; finally, they may be managing and recovering from floods. As locations where there have been either acute or chronic problems with flooding, it is hardly surprising that the case studies exhibited instances of these interactions. For example, in West Garforth interaction with the users of the playing field helped to secure this as a storm water storage location, while the West Garforth flooding group provided interaction and support for those recovering from a flood or anticipating a potential future one. A test of whether FRM is really being implemented comes when one asks about these interactions in relation to locations that have not been flooded in recent years.

In terms of flood prevention, the public have a stake in relation to land that could be used to locate sustainable drainage features, and their active understanding and cooperation is needed if standard practices are to change. The systematic promotion of sustainable drainage has the potential to remove significant quantities of flow from pressured legacy drainage systems and hence prevent or postpone investments. Such changes have been carried out in some Australian (e.g. Melbourne), European (e.g. Hamburg) and North American cities (e.g. Portland or Toronto), and there is potential to institute the same changes in the UK and elsewhere in Europe. Active recruitment of the public is needed, either as landowners in terms of their houses and gardens, or as local users of public land. To do so requires a significant education process in order to have public support rather than opposition for a systematic change to practices. It also requires planning and concerted action at a local level. In the USA, for example, Federal-level education processes (such as the US Environmental Protection promotion (EPA, 2016)) combine with examples of concerted local action, such as the work chronicled by Andrew Karvonen in Austin, Texas, and Seattle, Washington (Karvonen, 2011). In the UK, the importance of more sustainable approaches to drainage management is recognised, but they prove hard to implement in the context of fragmented ownership and responsibilities for drainage assets. Provisions in the 2010 Flood and Water Management Act to push forward the implementation of sustainable drainage on a more systematic level have not yet been implemented, and it increasingly seems will never be put in place due to resistance from development interests.

The public are also actual and potential victims of flooding in their homes, workplaces and in the public spaces they use. As well as the vulnerability identified on FRM maps of fluvial and coastal flood risk, many homes and businesses are potentially vulnerable to surface water flooding. Consequently, everyone, but certainly householders/workers in vulnerable places, should develop home decoration and management strategies that are ready for flooding to occur. A combination of property-level protection with emergency flood plans provides a significant route to achieve flood protection or to minimise flood damage and disruption. While the provision and receipt of flood warnings is well developed, these processes of local protection are relatively little practiced except in locations where recent flooding has occurred. As noted in relation to the Hull case, insurance processes mitigate against such measures at present. Moreover, as the Hull case also illustrated most vividly, where the public have suffered from flooding there is a need for joined-up systems of interaction that show compassion and care for those who are flooded through what is ultimately a very traumatic experience. The continued expectation that, when faced with a flood, a member of the public should be able to differentiate between different forms of flooding is an embarrassing legacy of a FD approach in

UK processes and procedures. Nevertheless, in other respects the UK is a leader in terms of the active involvement of flood victims in flood planning and management. The West Garforth planning process proved an example of this at the local level, but, at a national level, the National Flood Forum offers an example of an organisation that supports and campaigns in the interests of flood victims. Funding support for organisations like the National Flood Forum to provide a voice for flood victims when floods occur is an important and useful national investment. The ongoing and active involvement of civic society organisations at a local and national level has appeared as a route to maintaining and improving the quality of flood services in this respect.

Interestingly, when it comes to the issue of flood resilience, the case of the Netherlands is notable for its contrast with the UK. As Anna Wesselink (2007) stresses, levels of flood protection offered in the Netherlands are high compared to the UK – for example, offering a promise of protecting coastal areas from a one in 4,000-year flood. Nevertheless, Wesselink argues that this means at least 1 per cent of the Dutch population facing flooding once in a lifetime (ibid.). Moreover, the low-lying nature of the Netherlands means that 'if things do go wrong, in many places they are apt to go badly wrong', causing widespread flooding and significant risk to life (ibid.: 241). Added to this, risk calculations are based on ideas about dikes overtopping, when there is also risk of dike failure from soaking and undermining. Furthermore, climate change and the sinking of the Netherlands as the drainage policies dry it out suggest that, over time, the issues will only get worse. Overall, Wesselink diagnoses a case of political and technological lock-in to an idea of overall and absolute flood protection (2007). The impact of this is a complacent population that, despite considerable efforts to communicate ideas of shared responsibility by water managers, remains unwilling to believe that they suffer from flood risk and unlikely to undertake any flood protection measures (Kothuis, 2016). The consequence for the Dutch is a strange combination of high flood safety with the risk of catastrophe if floods do occur. Despite moving towards some aspects of FRM, the result is actually the typical technocratic extreme of high dependence of technological systems with catastrophic results if (when) systems fail. In this context, a UK approach that focuses on partnership, and the need to tolerate and manage around some flood risk, seems not only sensible (Wesselink, 2016), but significantly closer to the principles of a hydrosocial way to manage water.

This chapter has explored UK experiences of seeking to implement FRM. In significant contrast to the Netherlands, by combining the need to make space for water with the need to prepare for floods, the UK is aiming to steer a course towards something approximating to a hydrosocial flood management regime. When studied in detail, however, it is clear that there are ongoing significant challenges in implementing this approach. Not only do some members of the British public continue to expect absolute protection (e.g. those writing about Somerset farmers), but the UK government is failing to develop and support processes for managing surface water and for bridging from FRM into even closely related areas of activity, such as nature conservation. In conclusion, it seems that, while FRM's aspirations represent significant progress from the technocratic extreme of FD, implementation failures highlight the legacy of technocratic institutions that are unable to be flexible in the face of the complex combination of factors that contribute to the occurrence and experiences of flood.

8

FLOOD RISK GOVERNANCE

Introduction

Like water supply, protection from floods is, at least to some extent, a collective good: in other words, it is something that can appropriately be provided collectively to deliver mutual benefit to a group. Chapter 3 explored issues of water supply governance, considering the scale of water supply, the nature and extent of integration with other functions and also issues over who influenced the decisions. In this chapter, understandings of governance are extended to consider the approach to delivering collective goods in terms of governance culture. The understandings are exemplified using the example of flooding and the cases discussed in Chapter 7.

The aim of looking at governance culture is to explore the values and understandings about human society that underlie the way we undertake collective actions. This is important. By examining the implicit values and understandings of past actions, we can look with a better-informed and more critical eye at current policy. Originally developed as a means of exploring the different ways that societies deal with risk, the framework of governance cultures can be seen as a useful tool through which to consider how our society is approaching flood risk and other water-related risk. It relates to and expands the discussion in Chapter 3 through considering whether and how issues are linked together, and through exploring the assumptions implicit in the way that decisions are taken. Through utilising the governance culture framework to look at changing flood management in the UK, this chapter will demonstrate the usefulness of the framework in generating insights about any area of collective risk.

Governance cultures

The term 'governance culture' is derived from the 'cultural theory' developed by the anthropologist Mary Douglas (Douglas and Wildavski, 1982; Douglas, 1985, Douglas *et al.*, 2003) and utilised by Thompson (Thompson, 2008) and many others as a means of understanding how societies achieve collective action. Cultural theory identifies four archetypal governance cultures as a means to understand and catalogue struggles over the legitimacy of power between and within institutions (Tansey, 2004: 17). As depicted in Table 8.1, cultures differ in terms of the strength of allegiance to a particular group or institution – running from weak (left on Table 8.1) to strong (right), and the extent to which societal constraints circumscribe roles, experiences and knowledge – running from unconstrained (bottom of Table 8.1) to highly constrained (top).

The different governance cultures can be distinguished in a number of different ways. First, and most centrally, it is possible to identify the different extents/forms of

TABLE 8.1 Characteristics of the four archetypal governance cultures of 'cultural theory'

		Less allegiance to group/ community		*More allegiance to group/ community*
Constrained – more fixed rules	*FATALISTM*	*Collective governance*: Minimal *Assumed human nature*: Low levels of trust and cooperation *Input legitimacy*: People choose their own fate *Output legitimacy*: People can do what they want *Vulnerability*: individuals likely to organise where there are collective benefits	*HIERARCHICAL*	*Collective governance*: Rule-bound with designated roles and responsibilities in fixed silos *Assumed human nature*: Potentially fickle individuals give trust to institutions *Input legitimacy*: Only experts can assess the big picture *Output legitimacy*: Advised by experts, government can take decisive action *Vulnerability*: those in power do not know public preferences and their rules and silos stifle creativity and multifunctionality
Unconstrained – fewer fixed rules	*INDIVIDUALIST*	*Collective governance*: Goods and services shared through markets *Assumed human nature*: Individuals are self-serving, and individual/ organisational goals are selected according to the economic incentives they face *Input legitimacy*: Information conveyed via the market enables 'invisible hand' *Output legitimacy*: Can move quickly *Vulnerability*: A framework of rules is needed to ensure 'fair play'	*EGALITARIAN*	*Collective governance*: High levels of cooperation and participation across areas of responsibility *Assumed human nature*: Self-determining individuals oriented towards well being of the group. Flexibility in organisational goals *Input legitimacy*: Widespread consultation enables 'expert' input from everyone *Output legitimacy*: Decisions may take longer, but they are better *Vulnerability*: Participation takes resources and is potentially infinite, but ultimately decisions and actions are needed

Sources: Developed from Hood, 1998: 9, and Thompson, 2008: 21, 39

collective governance associated with each culture and, closely linked to this, the inherent assumptions about human nature: these are shown in the first two subdivisions of each cell in Table 8.1. Hence, for example, the minimal collective governance in the fatalist culture arises in a world in which human nature is assumed to be selfish and collective organisations are viewed with great suspicion. Each governance culture also claims to be legitimate, but the justifications through which they make these claims vary between cultures. This is depicted in the third and fourth sub-divisions of the cells on Table 8.1, with *input legitimacy* concerning the processes through which decisions have been made and *output legitimacy* referring to the extent to which a governance arrangement has the ability to 'get things done' (Van Kersbergen and Van Waarden, 2004). Hence, for example, a hierarchical culture claims input legitimacy because it draws on the best experts to consider what needs to be done for society; but it also claims output legitimacy because government departments with delineated responsibilities are able to take rapid and decisive action. Cultural theory also highlights that inherent contradictions in each culture drive an inevitable and dynamic process of shifting emphases through time: this is shown in terms of each culture's vulnerability on Table 8.1. For example, the instability and corruption caused by self-interest under an extremely individualist culture can take away from the smooth operation of trading in a company, prompting crises or scandal and a change of culture.

Each governance culture can be seen as an 'archetype', which should not be expected to be empirically observed in totality; rather, an observed culture may be described as including elements of two or more of the archetypes. Cultural theory does not attach a judgement about which of the cultural archetypes is better, but suggests, instead, that they are used for relative evaluations about how a particular institution has changed through time and/or to understand competition or tensions within or between institutions.

How do these cultures of governance fit with the ideas about governance discussed earlier in this book?

The Introduction and Chapter 1 explored the nature of technocratic water governance and the potential for a more hydrosocial regime in the future. As can be discerned from Table 8.1, many of the characteristics of an extreme technocratic decision-making approach correspond with what is described as 'hierarchical' culture. For example, it is an approach that puts a high value on experts making decisions that are appropriate for society as a whole. Likewise, in the call for more hydrosocial governance, the book is suggesting the need for a shift towards a more egalitarian approach. For example, the hydrosocial emphasis on supporting individuals to take responsibility for water management fits in with egalitarian assumptions about human nature: that people are self-determining but oriented towards the well being of the group.

These understandings were then expanded in Chapter 3's discussion of water supply history and governance. What Karen Bakker characterises as the 'hydraulic state' paradigm might be seen as a hierarchical approach, which emphasises technical decision-making and rules about access. As highlighted in Chapter 3, the shift towards 'market environmentalism' has operated at a range of levels, with different emphases in different countries. In terms of governance cultures, we can see this shift as encompassing a much greater emphasis on an individualist way of thinking, as water is exchanged in a market and the value of water-related ecosystem services are increasingly under discussion. With greater emphasis on stakeholder engagement, market environmentalism might also be argued to include elements of more egalitarian governance.

By discussing governance cultures, we can focus on the values underlying different approaches to water management. Manifestations of these values include but are not limited to: the ownership structures of water utilities, the way water services are priced and valued, the nature and content of regulation and the way the public is discussed and engaged. By speaking of governance culture we can avoid pre-judging an organisation or a regulatory regime. While discussing specific factors – for example, ownership of utilities – tends to close down debate by attaching stereotypes, discussing governance culture provides a means to question and explore the actions taken within a governance regime or by a specific organisation. Governance culture can act as a heuristic – a framework for classification – through which different approaches to water governance can be understood.

Flood defence and flood risk management

The different governance cultures can help us to understand some of the differences between flood management elements associated with flood defence (FD) and flood risk management (FRM), at least on a broad level. It should be noted, however, that in many policy areas the inherent values vary, depending on the way that policies are applied. In this respect, the broad stereotypes below should be taken as a 'first cut' of analysis, to be followed by the more detailed case-related discussion below.

In Table 8.2, the different measures for FD and FRM from Table 7.2 are considered in relation to the governance cultures.

As Table 8.2 illustrates, an emphasis on FD (associated with the first three measures only) to a large extent follows a hierarchical governance culture, in which government and experts decide what should be protected, when and how. Insofar as governments are not protecting individuals, we can read the approach to flood governance as more towards the left-hand part of Table 8.1: when insurance is available the culture is individualist, but if there is no insurance available then the governance culture is fatalist by default.

In contrast, FRM (encompassing all of the measures listed) provides multiple possibilities in terms of governance culture, depending on exactly how it is applied. In general, FRM is more oriented towards the collective well being of a region or catchment, rather than the narrower collective well being of a town or village that might be served from an FD approach. This arises because 'making space for water' implies a concern to pay attention to flood risk over a broader area than previously considered. The measures implied by 'making space for water' also include acknowledgement of the need to work beyond and outside the silos of hierarchical management

TABLE 8.2 Flood management measures in terms of governance culture

Element	*Governance culture and justification*
Construction of hard FDs	HIERARCHICAL: Generally draws on just local technical information and focuses defence on a relatively small scale (village or town)
Insurance against floods	INDIVIDUALIST: Insurance suggests a logic in which it is up to the individual to ensure that their home is properly insured against flooding and/or to choose a property that is not at risk. (It should be additionally noted that, in the UK, central government intervention in the flood risk insurance market ensures that even houses at risk from flooding can be insured. This involves a HIERARCHICAL compulsory subsidy from the non-flood vulnerable to the flood vulnerable)
Emergency services	HIERARCHICAL: The state provides emergency help
Making space for water	EGALITARIAN OR HIERARCHICAL: Involves drawing on both technical and cultural information about landscapes. It is more egalitarian if there is widespread discussion and working across siloed areas of management to decide where water should be stored, and how such storage locations can also serve multiple other functions. It is more hierarchical if technical experts dictate decisions, particularly if their considerations are focused on a narrow 'flood' goal without regard to wider considerations such as amenity or nature conservation
Resilience in building sites or design	HIERARCHICAL: If resilience is required by external organisations, e.g. through planning regulations INDIVDUALIST: If resilience is a requirement of insurance EGALITARIAN: If resilience is encouraged and supported, for example, through local authority support for a local flood group FATALIST: If resilience is left to individual householders to develop and pursue
Public knowledge of and preparedness for flood	HIERARCHICAL: If there are fixed rules about behaviour in the case of a flood warning EGALITARIAN: If resilience is encouraged and supported, for example, through local authority support for a local flood group INDIVIDUALIST: If there are tailored 'flood support packages' that organisations, individuals and households can buy into FATALIST: If resilience is left to individual householders to develop and pursue

and to pay attention to other local goals. In both of these respects, FRM is more 'egalitarian' than FD was. Nevertheless, the potential emphasis on individual household resilience and preparedness, particularly if these are not well-enabled by the state, creates a possibility that certain applications of FRM could be just a fig leaf to cover fatalist governance: pretty words that effectively leave each individual to manage their individual risk.

So which approach to cultures of FRM is better? As noted previously, cultural theory itself does not attach a normative value to the different cultures, instead arguing that they are just useful as an analytical framework. But we can see the historical story of addressing flood risk as having involved a shift through time. Initially, the system was fatalist: each individual or small community took what measures they could to avoid water and to eliminate their own risk. The development of the FD approach made systems more hierarchical; in some localities and circumstances, risks were considered sufficient for a governance body to take some collective protective action. With the shift towards FRM, the collectivity has broadened: not only are flood risks considered, but their interaction with economic and cultural well being and nature conservation is recognised to be important. Additionally, public perspectives are added to those of technical experts involved in assessing flood risk factors. Indeed, decision-making seeks to be holistic, taking a whole set of interconnected factors into account.

In the context of a book focusing on 'reconnecting with water' it should be no surprise that this shift towards more egalitarian governance is broadly welcomed. Although there are clear challenges associated with whether and how these different interests and expertise can be combined and lead to action, the aspiration that different factors be considered together and that the public have a say might be argued to be a way of rebalancing the technical focus of preceding hierarchical approaches to FD.

The flood management cases

Heuristics aimed an understanding governance culture are only useful if they inform our understanding of real situations. In the following sections, the discussion will consider each of the cases discussed in Chapter 7, but it will also consider which governance culture or cultures seem to have played a role in the activities undertaken. The purpose of such discussion is to develop a more politicised understanding of how flood management – or other collective action issues – are delivered and, hence, to open up possibilities and debate about whether and how they can be better delivered.

In York, a long history of flood management has meant that historic FDs have been supplemented in recent years by attempts to make space for water in the city, in terms of positioning of flood barriers to allow greater local storage, and in terms of aspirations for upland management far upstream from the city. Meanwhile, planning processes have imposed careful controls on new developments, considering the possibility of exceedance, and taking full account of flood risk and downstream implications. The approach does include many agency-oriented interactions with the public – for example, with farmers or with developers; however, there is limited evidence about an overall narrative about cooperation and partnership. In these respects, the flood management processes in York appear to continue to follow many characteristics of hierarchical – technically led – governance. Although the cross-sector working and interactions with the public introduce elements of an egalitarian approach, it is evident that this is not the dominant narrative. We can also note that, although protection against river flooding is a present and real issue in terms of York's water management, the extent of protection against surface water flooding – which is likely to arise at the same time as river flooding – is less well understood and the risks may be significant. A lack of political urgency in addressing this problem means that people have no way of knowing if their property is vulnerable to surface water flooding; in this way the approach is individualist.

In relation to West Garforth, it is useful to separate decision-making before the pilot project from the decision-making during and since its operation. We can see, in the period before the pilot project, decisions about FD were made by separate authorities – Yorkshire Water, Leeds City Council and the Environment Agency – according to their separate jurisdictions. The effect of this decision-making in West Garforth was that no action was taken and that flooding occurred. While the input decision-making can be seen as very hierarchical, the inability to get things done in this example was palpable. Specifically, although some properties were flooded every two years, they were too few in number and not sufficiently concentrated in a locality for any measures to be deemed suitable for funding. Hence, although the decision-making process looks hierarchical, the experience for the people of West Garforth was of a fatalist system in which no organisation saw them as worthy of attention.

The pilot project embraced what can be seen as a more egalitarian perspective on decision-making. Specifically, residents of the flooded properties interacted with technical experts from all relevant agencies in seeking to understand the nature of the flooding and in considering different viable solutions. Other community members were also consulted by the local authority in trying to drive forward the implementation of solutions in the aftermath of the study. Also, the cooperation created by the project provided the means to ensure some ongoing shared action on the problems of this suburb. Compared to the decision-making processes that existed previously, this new regime could be argued to be significantly more 'egalitarian' in terms of agencies cooperating with local people to discuss problems and solutions. The approach might be still more egalitarian if wider, non-flooded residents were part of a discussion about the potential for liberating parts of what are currently streets to provide upstream storage and local green space.

Before the 2013–14 floods, decision-making in Somerset might have been argued to be in transition from a hierarchical system to a more egalitarian one. The 'maintenance-of-agricultural-land-through-flood-defence' approach was being broadened to incorporate nature conservation interests and urban flood protection. The changes to the system were led by overarching shifts in the government funding regime from protecting agriculture to protecting homes. The new regime did not completely ignore agricultural interests, but it gave them less weight than had been the case previously, and their needs were balanced against those of urban residents and nature conservation.

The 2013–14 Somerset flood demonstrated the difficulties of achieving such a transition. Like the residents of West Garforth, the flooded residents of the Somerset Levels experienced fatalistic flood management; they felt that the flooding was a consequence of a flood management regime that ignored their interests. This is not a surprising reaction, given expectations formed under the previous hierarchical regime. The changes resulting from the flooding demonstrate the continuing greater influence of egalitarian approaches, in particular, through the promotion of upland storage of water, while also shifting to meet some of the farmers' expectations and aspirations in terms of dredging.

In advance of the 2007 floods, the Hull case provides a classic example of compartmentalised hierarchical decision-making. Each authority had clearly delineated responsibilities, but no authority had an overarching role, certainly not concerning the way that the different forms of flooding interacted with each other. In coordinating responses to the disaster, Hull City Council provided important and useful support and advice to many people, also drawing on the input and support of other agencies. As one might expect for an emergency response, the ordering and structuring of this work can be seen as hierarchical; while the regime could be caring, its focus was on particular categories of the 'most needy' and there was limited space for compassion in allocating resources to people outside these categories. Interestingly, the existence of the research project added an element of extra care, listening and feedback to the process – providing what might be seen as a fairly limited element of 'egalitarian' processes to the flood recovery experience.

Cultures of flood governance

Table 8.3 summarises the outcomes of the above discussions, expanding and developing the general point made earlier – that, with the shift to FRM, the explicit governance of flooding has changed from being largely hierarchical to egalitarian and hierarchical in combination. The term 'explicit' is important because it concerns the seen and talked-about aspects of governance, and is part of a number of caveats or complexities that the case study discussion has highlighted.

First, it is observable that the shift towards egalitarian governance associated with FRM involves a very broad range of activities, concerned essentially with 'taking everyone and everything into account'. In the case studies, 'more egalitarian' governance was signalled, both through cross-agency working (e.g. York) and through participatory processes with concerned publics (e.g. West Garforth). As illustrated by the Somerset case, however, the challenge of an egalitarian approach is the management of conflicting agendas. Identifying different agendas and why they conflict

TABLE 8.3 Summary of governance cultures demonstrated in the flooding cases

Case	*Hegemonic culture*	*Other cultural elements*
York	Hierarchical	Egalitarian and individualistic
West Garforth (before)	Hierarchical and fatalistic	
West Garforth	Egalitarian	Hierarchical
Somerset (pre-flooding)	Hierarchical and egalitarian	Fatalistic
Somerset (after)	Egalitarian	Hierarchical
Hull (pre-flooding)	Hierarchical	
Hull (recovery)	Hierarchical	Individualistic, egalitarian and fatalistic

aids analysis, but it does not always lead to clarity and consensus about the best communal route forward. In this sense, extreme egalitarianism could lead to stasis; alternatively, as in the Somerset case, it is accompanied with some eventual hierarchical decision-making.

Second, it is evident that the explicit governance culture is not the whole picture. Because they are the default 'non-action' (fatalist) or 'action by leaving it to the insurance market' (individualist) activities, the governance cultures to the left of Table 8.1 are implicit rather than explicit in the cases discussed. The analysis identified perceptions of fatalism/individualism when a flood occurred (for example, in the Hull and Somerset cases); it also flagged up the potential for such (non-)governance, as in the York case in relation to surface water flooding. The case of surface water management in York is particularly interesting because circumstances that York reports – with limited knowledge of assets and no resources to find more – are likely to be reproduced across most of the UK. Indeed, the fact that these issues are reported in York is indicative of the unusually high degree of water and flood management skills in York Council. Hence, it appears that while the UK does explicitly utilise hierarchical and egalitarian governance processes to manage the geographically and temporally acute risks of river and coastal flooding, the more chronic and diffuse risks, such as those arising from surface water management, are seen in less collective terms; this relative lack of governance is 'fatalist' by default. The on-the-ground impact is that many people who are not co-located are frequently flooded and our society treats this as their individual problem rather than a societal issue.

Third, for the authorities overseeing flooding, perceptions of fatalism – particularly when reported and perpetuated by the media – are the least desirable public reaction to acute flooding because such perceptions suggest that they are not doing their job properly. Minimising such public reaction after a flood hinges on the extent to which, first, it can be argued that all reasonable preventive and preparatory actions have been taken and, second, but closely connected, that the flooding event can be presented as something that could not reasonably have been predicted. This highlights how the shift towards egalitarian rather than hierarchical approaches has meant a need not just to join up the actions of different organisations, but also to ensure that the communications and messages are joined up and demonstrating cooperation. It also speaks to normative issues about which culture *should* oversee flooding. A hierarchical culture would work well if flooding issues could be neatly fitted into separate boxes. Because the management of different types of flooding overlap, and because flooding ties so closely into other issues like land use, however, it is arguable that some degree of egalitarian cooperation *is* necessary, otherwise the authorities will be seen as blaming each other in their communications when a problem occurs.

Fourth, and finally, the discussion has illustrated how, when we are dealing with the governance of risk, we are actually concerned with perceptions as well as the more tangible decisions about investments and land uses. We can recognise this because the governance culture (implicitly) attributed depends on who you are. In the Somerset case, for example, the change in regimes between 1985 and 2005 would be perceived differently by a conservationist and a farmer. The conservationist would be likely to emphasise the newly inclusive nature of the governance; hence, in the language of governance cultures, his/her perspective is that the regime has become more egalitarian. In contrast, a farmer in the same area would see the changes as a shift from hierarchical governance to fatalism: from the farmer's perspective the authorities have stopped listening to him/her and stopped prioritising those agricultural needs.

Conclusion

This chapter has introduced cultural theory as a means of analysing approaches to collective action. It has applied the theory to the general case of flood management in the UK, and then to the four flooding case studies discussed in Chapter 7. What, then, does the chapter argue about the usefulness of cultural theory?

In relation to flood management, Chapter 7 had already introduced the idea that the relatively narrow and siloed technical management of floods associated with the FD paradigm was shifting towards a more inclusive and joined-up approach to flooding that was called 'flood risk management'. The lens of cultural theory puts these claims in a political context – in other words, it helps us to see the changes that occur in terms of the two 'axes' of extent of collective orientation, on the one hand, and extent of flexibility, on the other. We can see FD as a relatively narrow form of hierarchical governance, which runs alongside fatalist governance of surface water management and other aspects of flooding that get no attention. In other words, it is a world in which flooding is viewed in a relatively fixed way – that is, either addressed by a clearly responsible part of government, informed by expert science, or it is up to the individual to look after themselves.

With the shift to FRM there is a big rhetorical change: the sort of governance the government is now claiming to provide is not just expert-driven, but also includes local people and pays attention to the things that cause floods and to the wider impacts that flood and measures to manage floods have – in particular, in relation to the environment. The rhetorical claim is towards a much more egalitarian form of governance – it is broader in scope, and both collective and flexible. The case studies, however, help to show what this involves, as well as the political tensions that it brings. First, as to some extent stressed in Chapter 7, change is not easy. It is notable that the communicative and inclusive changes associated with egalitarian governance have been slower to emerge (for example, in the York case) than the processes of joining up agendas. Second, egalitarian inclusion does not necessarily make everyone happy: in particular, it is likely to upset those whose interests were formerly the centre of the FD paradigm and whose role is downgraded through FRM, such as the Somerset farmers. Third, under tight resource constraints, some areas of FRM are likely to be neglected (meaning the default fatalist governance); typically these will be the chronic rather than the acute problems.

Overall, the concept of 'governance culture' enables one to examine 'How is the governance achieved?' in a particular situation. It offers a useful heuristic through which to understand changes and to chart differences between cases, or, potentially, between different parts of the water sector. As a means of analysis it goes considerably further than an analysis of ownership structures.

In a sense, the analysis of governance processes goes to the heart of this book. It asks whether shared goals are achieved through collective action or through individual action, and also whether these shared goals are achieved within fixed or flexible structures. It should be noted, however, that while the framework claims to be normatively neutral – not having a preference over what sort of governance is suitable – my use of it in this book is not. The position that has been argued since the Introduction is that water governance has been too technical and hierarchical in the past, and also a bit too inclined to see individualist market solutions as the answer to collective risk issues today. Instead, it is suggested, a more inclusive and joined-up approach would enable these issues to be addressed more fully and, potentially, also in a more efficient way. Hence, in relation to FRM, but also in relation to other water management issues, this book has argued for a rebalancing of technocratic water governance (influenced by hierarchical and individualistic approaches) in favour of a more egalitarian hydrosocial position, incorporating more joined-up thinking and more public engagement.

9

WATER IN THE LANDSCAPE

Introduction

Water has always been a crucial part of the landscape. Functionally, the needs for water supply, waste disposal and navigation, alongside the risk of flooding, have long shaped locational choices for settlements (Hoskins, 2005). However, water has also played an important aesthetic role; from landscape painting to holiday snaps, it offers a focus for our understanding of a landscape. Hence, no study of the relationship between the public and water could be complete without the consideration of landscape.

Within industrialised nations in the past, and more recently in emerging economies, urbanisation and industrialisation has led urban waterways to become working landscapes, and often also sewers. More recently, de-industrialisation in some locations has provided opportunities to reclaim water in the landscape for a set of multifunctional benefits, including aesthetic beauty, heritage conservation, biodiversity protection and flood management. But this emphasis on multifunctionality causes challenges for the funding and organisation of change: who makes the change in the organisation of the waterway and its banks, for what balance of ends and how is it funded? In this chapter I focus on four cases where active re-management of post-industrial waterways has taken place or is in prospect. While the cases are all drawn from the North of England, the questions they raise about responsibilities, ownership and action have parallels in other post-industrial landscapes across the world.

Like Chapter 4's focus on water in the home, this chapter has at its heart an interest in the different aspirations and management efforts that are directed towards water in a particular (landscape) scale. Rather than focusing on how water achieves a particular function, then, the emphasis is on how different functions are combined, or co-exist.

In terms of social science, this chapter extends the interest in collective action developed in the last chapter. However, whereas there the focus was on *what* is done collectively, here we also consider *how it is done*, considering specifically who is sponsoring/organising public engagement and how they achieve participation and mobilisation of public action. In this respect it expands on the discussion of mobilisation in Chapter 5 and on the nature of 'egalitarian' governance from Chapter 8, and hence explores some of the nitty gritty of relations between water service providers and the public, seeking to draw lessons to support a more socially oriented water management.

Public engagement – definitions and terminology

'Stakeholder engagement', 'public participation', 'customer involvement' and 'civic action' are some of the many terms used to refer to processes in which the public interact with a policy decision relating to the environment. They are applied most frequently to refer to the collation of public views in order to impact on a policy decision. However, the same terms are sometimes used to refer to processes when the public contribute to the implementation of a policy (though the latter are also sometimes called pro-environmental behaviour, sustainable lifestyles or sustainable consumption), a set of activities that we have already labelled as 'mobilising public action' in Chapter 5. As briefly explained in Chapter 1, here we will use the term 'public participation' to refer to all processes of involvement in *making* policy decisions, 'mobilising public action' to mean activity that helps to *implement* policy decisions and 'public engagement' as an umbrella term encompassing both sets of activities. Of course, the lines between different types of public engagement areas are not always clearly drawn, but range along a policy-action continuum with more participatory activities towards the policy end and more mobilisation-type activities towards the action end (Sharp and Connelly, 2002).

The organisations that might want to engage us might include government bodies, our water company, our local council, not for profit organisation, or multiple organisations working in combination. Invitations to engage may be open to everyone or focused on just a few; alternatively, processes of engagement may be 'claimed' with no invitation from the responsible body – for example, when a community group pushes for a policy change about a river bank, or carries out a litter pick on private land (Cornwall and Gaventa, 2000; Cornwall and Coelho, 2007). Each exercise has a sponsor – that is, the organisation that wants the engagement – and also an organiser, who may or may not be employed by the sponsor (Rowe and Frewer, 2005: 254–5).

Depending who is engaging us and what their purpose is, potential participants might be viewed as citizens, stakeholders, consumers, workers or residents, as members of a special interest group like anglers or mothers, or in a host of other identities. Whichever collective identity is ascribed, no set of people are ever completely of one mind, nor do they have a consistent set of needs, expectations, desires or abilities. A challenge for all engagement exercises is taking account of this diversity while still delivering the objectives required. It is useful to distinguish three overlapping understandings of potential participants in engagement processes. First, there is a mass of 'the public' viewed as a set of individuals or households (who, for example, live in an area, pass through an area, or are customers of a water company). They may all be invited to participate – but, depending on the issue, there may only be a minority who choose to do so. Second, there is an 'organised' version of the public, such as local community groups, religious groups, workplaces or other social institutions through which people can be accessed and which might enable some collective engagement. While working through such groups is often a short cut to accessing many people, it can miss out 'non-joiners' who do not choose to participate in any group. Finally, there is a set of professional stakeholders – working, for example, in civic society organisations, or in state agencies with overlapping agendas – who may also be appropriately engaged, particularly in participation (as opposed to mobilisation). Inviting participation from organised community groups and via professional stakeholders, such regulators can sometimes be seen as representing the public by proxy (Grunig and Hunt, 1984).

In giving new and greater priority to public engagement, the water sector is following the lead of many other areas of activity both in the public and private sectors (Arnstein, 1969; Grunig and Hunt, 1984; Healey, 1997; Cornwall and Gaventa, 2000). Rather than delegating decisions to experts and politicians, and receiving universal benefits, the consumer/citizen is regarded as having the capacity to actively make and shape the goods or services they receive (Cornwall and Gaventa, 2000). In terms of the cultures of governance, this can be seen as moving towards the lower 'unconstrained' part of Table 8.1 (i.e. individualist or egalitarian cultures).

For some theorists (Stirling, 2008; Wesselink *et al.*, 2011), there are four distinct rationales to processes of engagement. Engagements may be *instrumental*, if the aim is to diffuse conflict, to gain legitimation for a sponsor's policy or to achieve a particular action wanted by the sponsor. Many behaviourally oriented mobilisation activities fit the latter designation: they call for behaviour changes that benefit the sponsoring organisation. Participation processes that are tokenist – that is, ones that are run to be seen to happen rather than because of any inherent benefit – are equally instrumental. In contrast, engagement with a *substantive* rationale seeks to draw on the variety of other expertise to make better decisions or take better actions; the precise goals of decisions or nature of actions are therefore negotiable – hence there is at least an element of participation involved. The *normative* rationale believes in participation for its own sake as a desirable democratic end; it is likely to work towards the best 'ideal' type of participation. Finally, a *legalistic* rationale involves people because it is a requirement of law; it is arguably another form of instrumental engagement in which the objective is to avoid breaking the law.

Finally, in terms of linking this discussion of engagement to the earlier chapters, it is useful to note that a governance organisation's approach to engagement strategies will depend on the organisers' assumptions about society, and hence to the governance culture(s) that they believe applies (apply). Mobilisation processes are unlikely to occur in governance cultures with fixed rules and roles: if roles are fixed, the issue will be dealt with through normal processes and no additional action is needed. Hence, mobilisation is more likely to be associated with an individualist or egalitarian culture. In contrast, participation is unlikely to occur towards the left of Table 8.1; if everyone is focused on individual well being there would be no point in asking for their views about collective goals. Hence, participation initiatives are likely to arise from either a hierarchical or egalitarian culture.

Because this volume has not yet given it attention, the following section considers how social science theory can inform our understanding of participation – that is, the policy-making end of engagement. In the discussion of landscape management that follows, however, we switch between instances of public participation and public mobilisation, demonstrating some of the links and cross-overs between the two processes.

Public participation

Collating public views and knowledge to achieve 'public participation' in water-related decisions can offer some distinct substantive advantages to water managers. First, the public may have information that directly helps the water managers achieve their own watery objectives: for example, they may have information about historic uses of sites. Second, information about how the public use a landscape or amenity currently may be of relevance to future plans, suggesting, for example, how a feature should be designed in order to avoid wear and tear. Third, information about local public needs and desires which are not related to the sponsor's watery goals, but which nevertheless could be achieved by the planned action, may also be useful as it may identify how additional public benefits can be obtained, with the potential for reputational benefits to the sponsor and overall better value-for-money for society. Fourth, and finally, public participation has the potential to empower those involved, and perhaps to mobilise them to take greater ownership of the environmental assets in the future. In the first three advantages above the public are being treated as experts in their local area – able to contribute information of substantive relevance.

In addition to these substantive reasons, there are also instrumental, normative or legalistic rationales that might motivate sponsors to organise water-related participation (Wesselink *et al.*, 2011). For example, in the UK, a legalistic rationale is applicable for participation which is liable to create a substantial change in an area requiring planning permission. Some participatory exercises may also be 'instrumental', where the aim is merely to be seen to be undertaking the consultation and then getting the work done. Perhaps the least likely motivation for organising participation is a normative rationale; it is hard to see many circumstances in which sponsors can justify the costs of participation if it is not required, and if there is no wider substantive or instrumental benefit. In general, we might conclude that the egalitarian principle of 'taking decisions and actions together' suggests participatory processes justified and designed to achieve substantive ends.

Some of the substantive benefits that arise from public participation can be achieved with relative ease by just 'listening' to existing information and signals.

For example, some employees of a water utility will constantly encounter members of the public during their work and should be empowered to feed public perspectives into the organisation. Likewise, complaints or comments processes, as well as coverage in local news or online commentaries, provide evidence about public views that should not be ignored. However, all the benefits listed in the previous paragraph mean that there is also a strong case for issue-focused systematic collation of information about public understandings and preferences, particularly if a significant change is planned for public experiences or amenity (Reed, 2008). Early advocates of participation, such as Arnstein (1969), tended to argue that more participation is always better, but, in practice, any sponsoring organisation must balance the advantages of participation against acknowledged issues. Systematic collection of information via participation is both time- and resource-intensive and cannot be done for every decision; it is not always immediately clear which decisions warrant public participation and which do not. Moreover, if carried out badly, such 'systematic' public participation can have a negative effect on a sponsor's reputation and on a project.

Once a decision is taken to undertake a formal process involving the public, multiple questions follow concerning which 'public', when, where and how. In terms of the specific process, there are no universally 'right' answers to these questions, but careful advance planning is crucial. Published advice, such as Wilcox's *Guide to Public Participation* (1994), provides a good start. In his review of literature on stakeholder participation for environmental management projects, Reed highlights how participation should not be regarded as a one-off event, but rather part of a process with ongoing interactions which build trust between the sponsor and those participating (Reed, 2008: 2421). Drawing on the literature, he identified a number of axioms about the nature of best practice in public participation that are reproduced in Box 9.1. Even Reed's useful axioms, however, are built on the assumption that there is a clear distinction between experts who advise and support the sponsor and 'lay people' who form the public. As we will see in relation to engagement in water management, this is not universally the case.

Are formal processes of public participation always a 'good thing'? In her seminal discussion of the 'ladder of participation' Sherry Arnstein (1969) argued that they are, and that the greater degree of control that could be given to citizens, the better. Different types of participation were arranged as progressive steps on a ladder: hence, 'consultation' was better than 'information provision' and 'citizen control' was best of all. Since that time, Arnstein's influential, if idealistic, approach has been questioned for the assumptions that it is making about those who sponsor or organise public participation (Thornley, 1977; Wilcox, 1994; Sharp and Connelly, 2002). Frequently the scope for sponsors changing decisions is constrained by funding or timing; in the UK, for example, OFWAT's expectations mean that complete citizen control of water utilities' decisions is not an option. Wilcox's *Guide to Public Participation* offers a more nuanced approach, arguing that the right level of participation should be selected to meet the constraints of a particular decision, a point also emphasised in Reed's best practice axioms (Wilcox, 1994; Reed, 2008).

BOX 9.1 GOOD PRACTICE FOR STAKEHOLDER PARTICIPATION IN ENVIRONMENTAL MANAGEMENT

1. *Stakeholder participation needs to be underpinned by a philosophy that emphasises empowerment, equity, trust and learning.* This point highlights the need to provide appropriate information, to deal with power inequalities and to achieve two-way learning between participants.
2. *Where relevant, stakeholder participation should be considered as early as possible and throughout the process.* Reed argues that past projects have involved stakeholders in the decision-making phase, not in the project identification phase, when key decisions are made.
3. *Relevant stakeholders need to be analysed and represented systematically.* Reed highlights how stakeholder mapping or similar techniques are needed to ensure that a variety of those who have a stake in the problem are included.
4. Clear objectives for the participatory process need to be agreed among stakeholders at the outset.
5. Methods should be selected and tailored to the decision-making context, considering the objectives, type of participants and appropriate level of engagement.
6. Highly skilled facilitation is essential.
7. *Local and scientific knowledges should be integrated.* If scientific perspectives are a significant factor guiding the decision, then they need to be shared with the public as, in making a decision, you will need to reconcile the scientific and public views.
8. *Participation needs to be institutionalised.* The difficulties of achieving stakeholder participation often relate to the reluctance of the sponsoring organisation to commit to listening to and acting on an as-yet unknown set of comments.

Source: Reed, 2008: 2412–26

Academic commentators additionally highlight how participation processes are developed and used strategically (Sharp, 2002; Connelly and Richardson, 2004). Overall, we can conclude that participatory relationships with the public are useful and important in achieving legitimacy and substantively improving decision-making, but that they frequently take place in constrained circumstances, and hence should not be viewed uncritically.

Although it is important to recognise the differences between public participation and mobilising public action, the two types of public engagement cannot be completely separated. As has already been highlighted, both activities rely on trust between a sponsoring organisation and those who are (potentially) engaged. In this respect both types of engagement can be seen to form part of the ongoing relationships between water service providers and the people living within their jurisdiction, with additional parts of this relationship formed through complaints handling, customer relations and public relations. The links between these areas have important implications for water service providers; rather than employing communications professions merely for the purpose of reputation management, instead, all of these activities need to be regarded as linked processes which together could ensure that the public are both listened to (participation) and mobilised for the improvement of everyone's water system.

In the following subsections I look at four different case studies of landscape change and renewal to examine the potential and actual input of the public to these processes.

River Dearne restoration

This case tells the story of a stretch of river that runs through a semi-rural area within a mining district of South Yorkshire. It particularly focuses on how residents perceive a stretch of river that has been improved to support a better ecological habitat. While the 'restoration' of the river dated from 1995, the research it draws on, which was carried out by Emma Westling in 2009, demonstrates perceptions of the river long after the restoration had been bedded in (Westling, 2012; Westling *et al.*, 2014). The case is useful for two reasons: first, it helps to understand some of the recent history of landscape management; second, it begins to explore the factors that link to the relationship between the public and water – in this case a river – in their local landscape.

The case concerns the River Dearne, 10km to the west of Doncaster. It is a case study of what is known as 'river restoration' – that is, an attempt to intervene to improve the multifunctionality of the riverine environment. Like many such examples, restoration should not be understood in a literal way to relate to returning a landscape to a historically desirable norm, but rather about enhancing the landscape to achieve ecological and other benefits.

The case study site is a lowland landscape through which the original Dearne had meandered. By the 1970s, subsidence due to mining activity had disturbed the natural course, contributing further to flood risk from what was then quite contaminated water. These problems were addressed through the development of an artificial straightened channel somewhat to the south of the original river course. The closure of mines in the 1980s removed the chemical causes of pollution, but the straightened channel meant little variation in flow, inhibiting the development of wildlife habitats. To address the paucity of habitats in the river, in 1995 a 500m section of the river was 'restored'; stone barriers were placed inside the previously over-widened channel, creating a sinuous route with changes of bed and water velocity such that fish spawning could take place. The case is held up within the Environment Agency as a success story in which restoration has tangibly increased biodiversity. Nevertheless, for people visiting the landscape, these changes are not obvious unless they have prior knowledge.

At the time of the restoration, the process of impacting on a river was regarded in purely ecological terms, and not seen as something requiring local consultation. By exploring residents' views some 15 years after the restoration was completed, we are able to see their perspective on the final landscape created, rather than be over-influenced by recent works. An 'unrestored' stretch of the original straightened and over-widened channel runs immediately downstream from the restored reach, providing a form of 'before and after [restoration] comparison'. The work described here involved interviews with 16 residents living within 600m from the river, to understand how they viewed both the restored and original Dearne. The sample consisted of residents who self-selected from the 80 or so houses located nearby. The findings should not be seen as 'representative' of local views on the Dearne, but rather as indicative of how views are formed and how they vary.

Almost all interviewees saw this river as beautiful and tranquil, demonstrating that, despite its artificial channel, the river offered a valuable link with nature in their lives. They all valued its proximity and the access they could achieve from the paths along its banks. In discussing the river and its meaning for them, they referred to its current physical form and features they liked and disliked, but they also told stories of different encounters with the river, either from their own personal history or from accounts given by relatives.

All of the residents who chose to be interviewed had lived near the river during and since the restoration, and some also lived through the creation of the new channel in the 1970s. Many recalled the disturbance to the river as the 1990s restoration works took place, noting that it took a few years for the restored stretch to vegetate and start to look undisturbed.

While the local residents did not consistently show a preference for the restored or unrestored stretches, nevertheless, discussions revealed two distinct and somewhat contradictory perspectives on the river, with some interviewees voicing both perspectives alongside each other. The majority perspective had a preference for the unrestored stretch, which was seen as tidy, neat and cared for. This was preferred over the messiness characteristic of the restored stretch, which was seen as indicative of a lack of control. In particular, the vegetation in the restored stretch was understood to lead to accumulations of litter and hence signalled neglect.

The contrasting minority view preferred the restored stretch and its vegetation. This perspective demonstrated understanding of the tensions involved in river management: interviewees recognised that the more prolific vegetation might catch and hold litter and, likewise, that the provision of access also limited the extent of wildness around the restored stretch. However, the restored stretches were valued because they were felt to demonstrate more characteristics of a natural river.

There are several important factors that are highlighted by the Dearne case.

First, it is clear that past practices in river restoration paid little attention to local people and gave them a limited chance to influence decision-making. They were quite simply not regarded as relevant to the restoration. As the research showed, however, the restoration impacted on a landscape to which people had profound attachment and, from a current perspective, such lack of consultation looks like inept interference in an important part of their lives.

Second, while interviewees demonstrated good local knowledge of the history of the landscape, apart from such 'folk memory' there was no local means of finding this story. In terms of seeking to increase awareness and potential support for further restoration in the future, the provision of some local information about the restored stretch, perhaps in the context of wider information about the relocation of the channel from the 1970s, might help embed understandings of the landscape in those who visit and use it.

Finally, the research also reveals something of a challenge for the Environment Agency or any authority seeking to restore ecological health to rivers. Perspectives among the population vary, but it is clear that unhappiness about the messiness of vegetation and litter in the restored stretches is a significant and real concern, at least in the Dearne, and very probably in other 'restored' locations (see, for example, discussion of Malin Bridge below). If good-quality consultation had taken place it would have been necessary to listen to this concern and to take measures to address it. This would have cost more money. More optimistically, had consultation occurred, it is probable that local people would have had a better appreciation of why the stretch was being restored and what advantages the increased 'wildness' (messiness) provided. Perhaps some degree of local mobilisation could have contributed litter clearances and pruning of habitats, so that the local environment could have been maintained in a way that was valued even more highly by the local residents.

Five Weirs Walk

The Five Weirs Walk is another story that begins some 30 years ago, in the 1980s. It focused on the River Don, just downstream of Sheffield, in a part of the city that suffered profound de-industrialisation and dereliction in the 1980s. It is the story of a tenacious community group who took action to make the River Don more accessible to the public, and hence to embrace the amenity value of the river. The story is drawn from research carried out with the Urban Rivers and Sustainable Living Agendas Research Project (Sharp *et al.*, 2011a).

The Five Weirs Walk is a riverside walkway and cycleway that stretches 7.5km from Sheffield city centre, downstream to the eastern outskirts of the city. It was conceived, developed and steered to completion by a charitable trust, during a period of over 20 years. The walk was finally completed in 2007. The Five Weirs Walk can be seen as a key factor in the regeneration of the River Don corridor in Sheffield.

The Five Weirs Walk Trust was founded in 1987 with the aim of constructing and promoting a public walkway/cycleway along the Lower Don. The Trust involved a set of like-minded individuals with a shared vision. Though it is a 'community group', these people were largely knowledgeable and well-connected professionals drawn from the Sheffield Junior Chamber of Commerce and the City Wildlife Group. Many key individuals were involved for the entire duration of the project, making decisions, developing strategies and dealing with sponsors and landowners. Few, if any, of these people lived in locations adjacent to the proposed walkway (which was largely, but not entirely, a depopulated industrial or ex-industrial zone). There was an additional concept of 'membership' incorporating a broader and fluctuating set of supporters who volunteered to help with different tasks – such as vegetation clearance.

Two institutions were vital in the Five Weirs Walk gaining early support and success. Sheffield Development Corporation oversaw the planning and investment process in the Lower Don area from 1988–97; it underpinned the walk project by funding part of the early phase of construction. Second, the walk was written into the Unitary Development plan by Sheffield City Council, which was the relevant local planning authority, effectively ensuring that, where the planning authority was empowered to do so, it supported the development of the walk.

The Trust's operation is best understood through the means of two examples of their interactions.

The issue of the 'cobweb bridge' arose where a historic railway viaduct crossed the river. The Trust secured an innovative 'cobweb' design to carry the path under the viaduct, but was confounded in the attempt to build the cobweb bridge because they could not get access to the land. Land on both sides of the river was owned by a derelict timber yard. Directors at this point refused to negotiate. The Trust used its collective expertise to seek out a solution. Supporter lobbying and media coverage helped to secure the support of prominent organisations including Sheffield City Council and the Norfolk Estate. Eventually the owners of the derelict timber yard gave way and permitted the bridge to be built.

The Tinsley Alliance is the only example in the Trust's records of sustained collaboration between the Trust and a neighbourhood-based community group. Developers of the large out-of-town shopping complex at Meadowhall had undertaken to build a footbridge, which was essential both for the Five Weirs Walk and for the residents of the local area, Tinsley, to gain access to the transport hub and facilities at the shopping centre. The bridge had still not been constructed four years after the Meadowhall complex opened. However, the Five Weirs Walk and the Tinsley Forum joined forces and waged a successful campaign to get the bridge built.

In some respects, the Five Weirs Walk Trust can be seen as an example of governance for local people by local people. The idea of the walk was conceived and developed by individuals living in Sheffield. The group worked with location-based community groups when possible and appropriate – for example, in relation to the Tinsley alliance. Their goals and actions also fitted with the priorities of the democratically elected council. However, it is also notable that the Trust involved middle-class individuals from the west of the city asserting and organising to achieve their ambitions in relation to access to nature and leisure in the poorer eastern part of the city. They were able to be as effective as they were partly because they were connected through professional and friendship links with many powerful organisations across the city. Though community groups existed in the eastern part of the city, partnerships with them were not usually pursued; instead, the Trust's efforts were focused on addressing whatever immediate hurdles were faced in achieving construction of the next section of the walk.

In Arnstein's terms, the construction of the Five Weirs Walk could be conceived as 'citizen control', with a community group literally taking control and battling against inertia to make change happen. It is 'claimed' participation rather than invited participation; nobody asked the group for their comments on a riverside walk – it is the Trust that identified the need and successfully argued for it to be on the political agenda. But, from another perspective, the Five Weirs Walk Trust's members are the sponsors of change who impose their ideas on the local area with relatively little in the way of consultation or local participation. The Trust would be likely to respond that they were caring for a relatively de-populated former industrial area, for which nobody cared; but because they did not try to organise wider participation we cannot say this with complete confidence. Ultimately, the Five Weirs Walk is a very good news story of community involvement in making a watery landscape accessible to local people. It is a model, however, that should be treated with care. In a different landscape it might be a case of one powerful set of citizens imposing their vision on a river and a set of citizens for whom it is not the right thing.

The Five Weirs Walk project involved mobilisation – of the wider membership to work on litter picks and other activities, but more crucially 'self-mobilisation' of the committed group who formed the vision and pushed the plans forward over a 20-year period. This represents a significant personal investment of time and energy from a set of committed individuals contributing to the public well being.

The example raises issues about the geographical scale at which 'community' should be conceived: is the Five Weirs Walk Trust a 'community' group because its leading members come from Sheffield, or is it a set of outsiders because its leaders do not come from the residential areas adjacent to the proposed walk? The example also highlights the difficulty of seeing stakeholders as neatly divided between professional experts and the lay public. In this example – and in others, as we will see – community groups include individuals who are highly expert in a variety of ways. In practice, whether we are dealing with individuals who are part of a professional organisation, a community group, or neither of the above, what we have is a continuum of individuals who are more or less well connected in certain fields, and also more or less informed about contemporary relevant issues.

Malin Bridge flood alleviation

The Malin Bridge site is an area of former industrial activity at the confluence of the Rivelin and Loxley rivers a few miles north west of Sheffield city centre. When, in 2007, severe summer floods hit Sheffield, the area's vegetation was the source of woody material that blocked downstream bridges, contributing to the flooding. In their post-flood study, the Environment Agency identified the area as in need of vegetation clearance in order to relieve potential future flood risk. The work described here draws on research by Peggy Haughton investigating local experiences of flood clearances (Haughton, 2015). It is an interesting case study to discuss because the perspectives voiced demonstrate the nature and strengths of people's attachment to the landscape around the rivers. In this example, technical experts from the Environment Agency were prioritising flood protection at the potential expense of nature conservation, which many local residents sought to protect.

Sheffield's industrial take-off and its fame as a centre for cutlery manufacture was underpinned by fast-flowing local rivers, which drove hammers and grinding wheels. Its early industry was located in the river valleys upstream and west of the city centre; but these were abandoned in the late nineteenth and early twentieth centuries, when, first, steam and, later, electricity provided more flexible and footloose sources of power. In a reversal of their former role, the river valleys were abandoned and vegetation took over from the industrial workings, forming a series of green tongues of parks and woodland reaching from the local uplands into the city. The Malin Bridge site, at the confluence of the Rivelin and the Loxley, is one of these sites, well known locally for being at the junction between different routes out of the city and also for being located beside bus and tram stops. Perhaps because of its accessibility to public transport, the Malin Bridge area was widely used from the 1920s onwards as a gateway into the greenery of the Rivelin and Loxley valleys. Its wooded nature led a local school to declare the area a 'nature reserve' in the 1990s, while ecological surveying just before the clearances revealed that the area was a good habitat and that the water quality (and associated aquatic species) were also high. The site also attracted negative attention, however: the perception that it was neglected had led to the organisation of at least one local litter pick in 2003.

In June 2007, heavy rain on saturated soil led to severe flooding in Sheffield, affecting 1,200 homes and 1,000 businesses (Environment Agency, 2007). While many of these were downstream of the city, upstream the River Loxley burst its banks just below Malin Bridge, flooding a number of business premises, roads, public transport facilities and also some sheltered accommodation. The fast-flowing water also passed through the area of dense foliage around the bridge, picking up wood and debris that it carried downstream and was subsequently seen as responsible for causing some other flooding.

Immediately after the floods, remote sensing identified the Malin Bridge site as one of a number of locations where channel and riparian vegetation could be managed to reduce potential for blockages and associated future flood risk. As an Environment Agency operations manager described it: 'we've spent a lot of money and time after the floods in 2007 picking trees out of bridges, and really it's easier to pick trees out of the ground, rather than pulling them out of bridges' (Haughton, 2015: 144). Monies were available from Defra to carry out such actions, and a process of planning and consultation built up to works being carried out in late summer 2009, a timing selected to minimise the impact on birds and fish, but argued later to have led a rushed process of consultation. The consultation involved three major components. The council was consulted at a programme level about the combined plan for channel clearances through Sheffield, while local consultation aimed to contact landowners with a material interest in the specific works, as well as a selection of local interest groups. In addition, local interpretation boards were put in place adjacent to the site.

Subsequent criticisms of the consultation process focused on who was involved, but also on the tone and agenda of the interaction. Erroneously, the consultation had excluded some riparian owners whose land was directly affected by the works. Some residents also found the description of the site as 'Malin Bridge' confusing; their perception from the consultation was that work would take place immediately around the bridge, not some distance up river as actually proved to the case. The selection of contacted groups was also criticised as being too narrow; however, with the site lying at the border of administrative boundaries, the identity of potentially concerned groups may not have been immediately obvious.

The tone of the consultation was largely informative: Environment Agency experts said that clearance was necessary to protect people from flooding, and the works were what was needed to achieve this. An interviewee from a member of a consulted group indicated that he had understood the extent of the planned works, but that it was hard to object when you were told that such works were necessary to protect people downstream from flooding. In other interviews some people questioned the Environment Agency's confident claims about the necessity of the work. For one commentator from Sheffield Council, the work constituted a 'very sort of engineered focused programme [of flood management]' (ibid.: 159) – implicitly viewing the work as utilising a traditional flood defence (FD) approach as opposed to more contemporary catchment-based soft solutions that are associated with flood risk management (FRM). Other comments from council workers indicate that points they made about the site were not taken on board. Concerns were also raised about an ecological report on the site, available within the Environment Agency but only released to stakeholders after the work had commenced. The interpretation boards were also much criticised for giving little information about what was planned. They included some photographs of the site in the past, but did not explain what works would be carried out. A consequence was that many residents and regular passers-by were shocked when the works commenced.

During August and September 2009, 400 tonnes of trees and vegetation were removed from the site, alongside 4,500 tonnes of earth and gravel. The work excited much local interest, and people collected along the bridge to watch the activity that began on its western side and stretched further upstream as it developed. The excavations revealed an old weir and also a millpond, somewhat altering the route of the channel. Some of the confusion noted above about the scope of the work arose because the site expanded somewhat while the contractors were on site: 'Really we were supposed to stop at the weir, but we'd taken some trees out, and then you can see up the weir, and then there were trees that were problematic and, you know, well we're on site with all the machinery here' (ibid.: 177). Piles of material remained on the site until October 2010, when full access to the site was completed and the material removed.

The impact on the local area was visually dramatic. What had previously been dense woodland became cleared, revealing the historic weir and millpond. Views about the appropriateness of the final state of the site were varied. Some very much favoured the new appearance of the site. It was seen as being returned to its 'original' (pre-vegetated) state, and as enabling the confluence of the rivers and the historic waterworks to be viewed. It was also perceived as tidier and more manageable, providing improved sight lines for traffic and pedestrians. In contrast, others were very upset by the loss of greenery and ecology associated with the change. 'I just think it looks really ugly now' (ibid.: 156), commented one interviewee, while others mourned the loss of wildlife. The latter was confirmed by an independent ecological survey that showed dramatic decreases in biotic quality between the summer and autumn. A particular ecological concern was the re-exposure of the weir, which was seen as constituting a new barrier to fish.

Though these two types of responses can be seen as polar opposites in terms of their preferences for how the site should be, they also had certain similarities. People on both sides of the debate related passionately to the river and the surrounding landscape, and expressed their views on it in strong terms, even some months after the change. Though highly vegetated and at some distance from local facilities, the site clearly has an important role in many people's lives. Also, in an echo of the findings about the Dearne, for both 'sides' the site was actively understood through memories, whether these were direct individual memories, other people's anecdotes, or collective histories constructed from photographs or other sources.

So, how should the story about the Malin Bridge flood clearances be viewed?

One part of this story is the process of inept and tokenist consultation carried out by a technological agency anxious to be seen to have taken action following a flood. Judging from the way it was conducted, the rationales were instrumental and legalistic rather than substantive. In this respect, it might be seen as a replay of the way that technocratic water management has long been assumed to interact with the lay public. And, indeed, it is very interesting that as late as 2009 the Environment Agency were still conducting consultation in this way. However, there are layers of the story that are different from the stereotype of technocratic–lay interactions. First, like in the case of the Five Weirs Walk, the public were far from inexpert. Indeed, members of the public's expertise equalled or bettered the Environment Agency and the Local Authority in terms of their knowledge of site history, its current uses and its ecology. Second, the flood management authorities were not in agreement about this issue; the Environment Agency (at least the flood management part of it responsible for the clearances) had one perspective and Sheffield City Council had another. Third, the relationship between flood management and ecological care is interesting to ponder. For the Environment Agency (and for the traditional, 'take it back to how it was 100 years ago' narrative from local people), vegetation was inimical to flood management. However, as the Sheffield Council interviewee pointed out, this is a very old-fashioned perspective on flood management, contrasting with contemporary emphases on catchment management in which ecology is seen as an ally not a foe in the struggle to manage flood waters. In this respect the clearances can be seen as an FD response, while maintaining and managing the vegetation might have been closer to an FRM approach. There is clearly a debate to be had here, and a need to balance some different flood and non-flood interests in a specific context. In this example the debate was curtailed through an inept public consultation process coupled with temporal imperatives arising from the bird-breeding season alongside the need to be seen to have taken action in Sheffield following the 2007 floods. It seems a shame that, in the Malin Bridge example, this opportunity was lost, and we are left to speculate on whether and how ecology, history and short- and long-term flood mitigation could best be reconciled.

Bradford Beck

As discussed in Chapter 6, Bradford Beck is a highly modified watercourse with a number of tributaries draining an intensely urbanised basin of about 60km^2, centred on the northern English city of Bradford. Chapter 6 focused on specific measures that are underway to address the quality of the water in the beck. Here, the concern is less the in-channel ecology and more the institutional arrangements; specifically, the section addresses who is taking what action and how those relate to public understandings of the beck within their local landscape.

As outlined previously, the beck was channelised and culverted during the Victorian era, when it acted as a conduit for waste water. The beck was cleaned but diminished by the development of a sewer network, and then became still cleaner with the loss of Bradford's remaining heavy industry in the early 1990s. Despite these improvements in its water quality, the beck and its tributaries have been left, hidden, culverted and largely forgotten, snaking through the back streets of the city. In this respect it is typical of many urban watercourses today (de Graaf, 2009; Karvonen, 2011; Cettner, 2012). In common with a minority of these urban watercourses across the globe, the 'luck' for Bradford Beck is that a group called the Aire Rivers Trust has emerged and decided to focus on the beck as a locality for action. The objective of this section is to discuss how the group worked to develop a vision for the beck, and to explore what potential they have to realise it. The story told has parallels in community mobilisation efforts made in other locations, such as the activities chronicled by Karvonen (2011) at Country Club Creek in Austin and Longfellow Creek in Seattle.

The Aire Rivers Trust was set up in 2011 by a group of people concerned about the quality of the water in the River Aire. It is called the Aire Rivers Trust in recognition that paying attention to the quality of the Aire meant addressing water quality in its tributaries, including Bradford Beck (in the same fashion, the daughter, group concerned with the beck, calls itself Friends of Bradford Becks (FOBB)). The Aire Rivers Trust follows in the wake of earlier campaigns such the 'Eye on the Aire' of the mid-1990s. As the groups' commentary on the Rivers Trust website indicates, the Trust aims to help in the process of improving the Aire and to ensure that the public recognises and benefits from such improvements (Rivers Trust, 2014). Activities undertaken to achieve this aim include improving fish passage, river clean-ups, educational projects, water quality monitoring, improved access and eradication of alien invasive plant species. A key success for the group was securing funding from the Environment Agency to lead a pilot process for developing a catchment management plan for Bradford Beck in 2011–12. The funding paid for a worker to interact with stakeholders over the course of 12 months, enabling the development of the plan. The Environment Agency additionally funded some days of professional facilitators to run public meetings or other consultative events.

The catchment management plan process consisted of a combination of stakeholder interaction and action. Consultation and negotiation was carried out with relevant local groups and organisations, including Bradford Council, the Environment Agency, Yorkshire Water, large local landowners/employers such as Bradford University and other environmental groups working in the district (Aire Rivers Trust, 2012). Some public consultation events also took place, supported by the professional facilitator. Publicised among existing contacts and by word of mouth, these events were, to a large extent, attended by those who had had previous interaction with the Aire Rivers Trust, especially those linked to the public and voluntary sector, as well as members of the public who were linked to these sectors. In this respect they effectively accessed some of society's 'joiners' (people who choose to join local groups and are connected in their neighbourhoods), but did little to reach beyond them to the 'non-joiners'. The public meetings provided space for participants to indicate the ways in which the beck was important to them, and also to highlight activities already underway that impacted on the beck.

The format of these public consultation events collated views in a highly structured way; a cynic might conclude that the purpose was instrumental, to gain legitimacy for the goal of the group and their plans. To a substantial extent this is probably correct. However, the events also allowed space for debate and provided room for views and preferences to be expressed. A further important consideration is that the plans were not controversial – it seems improbable that any person would want to argue that the beck should stay polluted, for example. Overall, the event should be viewed as part of the wider discussion with organisations and groups in the area to draw perspectives together and find substantive ways to make a difference to the becks.

The final catchment management plan identified a vision of the future Bradford Becks as clean, visible, accessible and ecologically cared-for rivers in the context of a city that used and disposed of water wisely (ibid.). The plan specified 12 actions that were couched as steps towards achieving this vision. Some of these actions can be seen as signature projects – that is, work-intensive projects with tangible outcomes are central to the Trust achieving its aims for the beck. These included a pilot renaturalisation (or 'restoration') of one part of the beck, the improvement of water quality over the whole beck and the development of an FOBB group. Another set of actions can be seen as relating to the beck, but already underway, motivated by other factors. These include Bradford Council's work developing an FRM plan, some of the activities undertaken by Yorkshire Water to address pollution, and the link between the route of the beck and proposed cycle-ways. A final set of actions are concerned with engagement, publicity and awareness. Some of these are quick wins – for example, replacing signs saying 'contaminated water' with the name of the tributary beck, litter picks, clearances of invasive species and also the development of nature trails. Others involve more targeted interaction with particular sets of stakeholders – for example, education and support for riparian owners, or a challenging set of negotiations, for example, to find a way to mark on Bradford streets the locations where the beck flows beneath.

In isolation, none of the actions listed would have a significant effect. However, by cleverly weaving together these three different types of actions, the catchment management plan might be argued to have created a plan that is more than the sum of its parts. By including a number of beck-related actions already underway the plan achieves an 'easy win'; the feel-good publicity of the link to the beck is provided for partner organisations, but at the same time the partner knows that their activities are being watched by FOBB and that they will be pulled up short if they fail to deliver promised actions. The publicity and engagement activities are also crucial; they usually involve 'feel-good' elements, either for the individual (e.g. enjoying a nature trail) or for the perceived benefit of the beck (e.g. a litter pick), but they also offer a means to publicise the existence of the becks and to demonstrate that they are cared for. In this way, the activities have an important cumulative impact in showing, even to 'non-joiners', that their locality has the benefit of a river. Finally, the signature activities offer the substance to enable FOBB's leading individuals to feel that their efforts are achieving tangible impacts. For example, if the group achieve the renaturalisation of a section of the beck then this not only makes a (small) tangible different to the beck, but it will also symbolise and benefit from all of the other activities undertaken with the management plan.

On the surface, the Bradford Becks catchment management plan and the associated FOBB Facebook page and websites are the epitome of localism. Here, local people have risen up in large numbers (those signed up to the Facebook site now top 300) to support action to improve their local rivers. In practice, like the Five Weirs Walk group 20 years before, the group is led by a relatively small group of motivated and dedicated volunteers who invest time and significant professional expertise in effecting the changes listed. It is also notable – and to the credit of government and the Environment Agency – that the case is a form of institution-sponsored localism. The group were given a significant boost through securing funding for a worker to support the development of the catchment management plan. Many of their further aspirations involve securing more money from the government or other charitable bodies. Whether the vision – or something like it – is realised will depend on the group's ability to maintain this momentum and action – that is, to motivate its key volunteers, to offer tangible change and activities for the public and to marshal the actions and funding of institutional stakeholders.

The story told above does not *yet* demonstrate significant changes in the policy or activities of any of the key governance organisations – the Environment Agency, Bradford Council or Yorkshire Water. In this respect we might say that the FOBB have not yet been successful at achieving 'participation' in policy, defined at the beginning of this chapter as public involvement in *making* policy decisions. In practice, since the initial draft of the material was written it has become clear that Yorkshire Water has sought and won extra funding from the regulator, OFWAT, to address some of the pollution issues in the beck (Lerner, 2015). And other changes in policy may follow. In this respect the important point is that the group has itself used processes of public participation to mobilise new individual and collective actions alongside existing formal institutional activities in the apparent direction of one cause. As a formula for achieving change it looks like it has the potential to drag some formerly rather reluctant institutions together to jointly address a local issue.

The Aire Rivers Trust is a hybrid case: like the Five Weirs Walk, there is an element of 'claimed participation' because they have identified and pursued an issue about Bradford Becks which was not previously on the political agenda. However, the sponsorship from the Environment Agency means that, on a broader national basis, we might see this as an example of 'invited' mobilisation.

In terms of the Trust's own processes of interaction with local organisations, we can see the overall process as having a substantive rationale in terms of mobilising local organisations and people to rally around the agenda of improving Bradford Becks and play a variety of different parts in trying to make a change happen. While the specific public consultation meetings may be seen as more instrumental, they form part of a wider substantive process of mobilising an alliance of individuals and organisations to achieve change for the public benefit.

Conclusion

As the research stories told in this chapter tangibly illustrate, water in our landscape is shared and generates strong feelings of attachment. The strengths of people's collective responses are shown by how interviewees talked to researchers about their local water, but also by the time and energy invested in improving the rivers. Water is valued in the landscape both because it is enjoyed aesthetically, providing a link to nature, and because it offers a focus through which the history of an area can be understood. People's strength of attachment to rivers, also highlighted in numerous other works and expressed in relation to other bodies of water, is important and may provide a route to support the further improvement our landscape; it could also be drawn upon to help influence other public practices impacting on water (Strang, 2004; Brown and Clarke, 2007; Karvonen, 2011, Woelfle-Erskine, 2015).

Because people value water highly, access to water is prized. This was clear both in the cases where access was already available (Dearne and Malin Bridge) and by the Bradford Beck and Five Weirs Walk cases, where achieving and promoting public access formed a shared agenda around which mobilisation occurred.

Beyond the consensus on importance and access, the cases illustrated some divergent perspectives on rivers. Specifically, in both the Dearne study and the investigation of flood alleviation at Malin Bridge public perspectives on the rivers showed mixed reactions to ecological strength and diversity, with one view favouring neatness, tidiness and controllability while the other favoured greater naturalness. To some extent these divergent public views parallel the change in professionals' approach to the management of water. Traditional technocratic approaches that seek to control the water demonstrate a similar mind-set to those members of the public who want a tidy riverside. More contemporary approaches that emphasise the need to work with nature may be closer to the public mind-set that values more natural rivers, even when it leads to 'mess'. What is not clear from these other cases, or from other work in the literature, however, is whether, when and how these divergent perspectives can be reconciled.

On the face of it, each case includes both institutional experts and 'lay' stakeholders. In a technocratic perspective, the institutional experts would hold the monopoly of technological knowledge while lay people would experience emotional responses to the water. Likewise, a technocratic view would anticipate that institutional experts would initiate, plan and implement proposals for change, with lay stakeholders playing a bit role influencing aesthetics and the details of plans during a 'consultation' stage. The cases question both of these assumptions.

First, as all cases showed, expertise is not confined to those in formal institutions, nor emotional responses to the lay public. Almost all 'lay' people in a locality are more expert than institutional stakeholders in relation to the way that the area is used; after all, they live locally and use the area so have the ability to observe usage regularly outside of working hours (this finding has also been stressed in other studies of expertise – for example, see Collins and Evans, 2002). Moreover, as these cases show, many lay people are also experts in a range of technical matters such as ecology, engineering, or the identity of locally relevant stakeholders. In fact, many of these volunteers take the role of institutional expert in another location/sphere as their day job. Likewise, just as institutional stakeholders do not have a monopoly on expertise, so the cases also show how lay stakeholders are not the only people to feel an emotional attachment to place. As was particularly well illustrated within the Malin Bridge example, both council workers and Environment Agency staff felt passionately about how the river should be.

Second, the processes of initiating, planning and implementing actions were undertaken both by the institutional and civil society organisations. Within the environmental literature, civil society organisations are most often discussed as protest groups, arising in response to damaging aspects of industrial society and leading public concerns about the activities (e.g. Walker *et al.*, 2010; Cotton, 2016). With the exception of the large and long-established environmental organisations such as the UK's National Trust (Lowe and Goyder, 1983), the visibility of civil society organisations in initiating and leading environmental change is limited. However, as the examples in this chapter clearly highlight, small and single-issue environmental groups can and do mobilise action on an issue and can achieve significant changes within a local environment. The Five Weirs Walk Trust and the Aire Rivers Trust both 'claimed' participation in relation to rivers, and they both mobilised and coordinated action to help change to occur. Notably, the issues that they were seeking to affect were not the designated responsibility of any organisation, but fell between remits. Likewise, they were not controversial issues with strong opposing interests, but rather the lack of previous action related to inertia, alongside a lack of larger-scale political will to invest in the quality of water bodies and watersides.

Third, in relation to participation, while some of the examples do play to the technocratic stereotype of a designated expert body failing to consult, or consulting badly in a way that is legalistic or instrumental, there are also significant counter-examples where either the designated authority is more participative than this model would predict, or where the civil society organisation demonstrated technocratic tendencies. Examples fulfilling the technocratic stereotype include the involvement of Bradford Council and Yorkshire Water in the Bradford Beck process. Likewise, the complete lack of consultation on the Dearne restoration, and the rushed and instrumental consultation about the flood clearances at Malin Bridge demonstrate the Environment Agency fulfilling a technocratic stereotype. However, it is the same Environment Agency that offered the funding that enabled the Bradford Becks work to be developed. Moreover, the Five Weirs Walk Trust used very little participation in pursuing its clear agenda. These complexities seem to illustrate the relative unimportance of the formal position of organisations sponsoring participation; rather we need to look in detail at what is done, how and with what effects before reaching judgements about engagement processes.

In the context of the wider narratives of this book, the Dearne and Malin Bridge cases provide examples of technocratic governance in which public engagement is non-existent or tokenistic. In contrast, the examples of Five Weirs Walk and Bradford Beck show active community involvement not only in improving the landscape, but also in demanding and driving changes in policy to achieve that end. In this respect, it is not surprising that decision processes in both cases exhibit elements of egalitarian governance (see Table 9.1). Interestingly, both groups obtained support from governance institutions (Sheffield City Council for the Five Weirs Walk, the Environment Agency for FOBB) to push forward a progressive agenda.

While the examples in this chapter do not all provide exemplar cases of how interactions between water authorities and the public should look within a more 'connected' form of water management, they do point to some principles about how such water management might be organised beyond (just) public interactions with the landscape:

1. People care about water, particularly water in the landscape. Hence, public engagement around water issues is possible and can be substantively useful in the design and implementation of water actions. Water-related engagement can enable water goals to be achieved more cheaply, to a greater extent, or to deliver co-benefits meeting other societal aspirations.
2. While important processes of engagement may be invited by water authorities, useful forms of engagement will also be 'claimed' by civic groups, making good perceived policy failures or blindness – for example, in relation to urban rivers. In this respect, civic organisation can have the chutzpah to develop new 'civic imaginaries' (ideas about how the city can be better or different) that Andrew Karvonen has called for (2011).

TABLE 9.1 The Five Weirs Walk and Bradford Becks decision processes as egalitarian governance

Egalitarian characteristics	*Five Weirs Walk*		*Bradford Becks*	
Collective governance: High levels of cooperation and participation across areas of responsibility	Close cooperation with council and with Sheffield Development Trust	H	Supported by Environment Agency. Activities stimulated the involvement of Bradford Council and Yorkshire Water	H
Assumed human nature: Self-determining individuals oriented towards well being of the group. Flexibility in organisational goals	Individuals running organisation were fighting for what they perceived as a collective good	H	Individuals running organisation were fighting for what they perceived as a collective good	H
Input legitimacy: Widespread consultation enables 'expert' input from everyone	Little consultation about the goals of the group	L	Participatory exercises were highly managed, but enabled widespread input	M
Output legitimacy: Decisions may take longer, but they are better	They have achieved their goal	H	They have made a splash in the press but their impact on the becks so far is negligible	M

Note: Characteristics taken from Table 8.1 and rated as H (high), M (medium) or L (low) for each case

3. A variety of 'expertise' (concerning, for example, chemistry, ecology, local uses, history and flood management) is likely to be relevant to any water issue. This expertise is spread through society, located variously in governmental, private, civic sectors and among the non-joining public. The implication is that expertise within those inviting participation is partial and needs to be so-perceived if the benefits of others' expertise are going to be appreciated.
4. Organisations with focused goals will sometimes invite participation for only substantive or legalistic reasons, potentially missing out on the opportunity to address water issues from a broader perspective, drawing on more and more varied expertise and losing opportunities to achieve their own goals in a more sustainable and multifunctional manner. Hence, while ambitious strategies and goals are good, they require alliances for action to be built and for some flexibility about goals and outcomes to accommodate varied interests.
5. Public engagement can help develop and explore issues but it cannot overcome value differences in society. Hence, for example, higher-quality public engagement about the vegetation of Malin Bridge might have identified contrasting imperatives for ecological maintenance and local flood risk mitigation, but it would not have absolved authorities from reaching a decision on where and how these requirements might be balanced.

Overall, this discussion of the role of water in the landscape raises a sentiment of optimism and positivity in relation to water management. It is clear that people care about water, and people and institutions can – sometimes – collaborate to achieve a form of water management that is more multifunctional and does serve to support different needs, on some occasions. The question that this chapter raises for the rest of the book, however, is how this sense of possibility can be infused through the world of water management such that it is more systematically embedded in the way institutions and citizens interact to examine, plan and manage our use of this most important resource.

10

CONCLUSION

Linton (2014) has indicated that a focus purely on the physical features in the hydrological cycle is giving way to a new 'hydrosocial' approach to water management, concerned not just with the physical processes of water management, but also equally with the humans who extract, utilise, treat, manage, guard against and dispose of it. Somewhat less optimistic than Linton, in this book I have suggested that while rhetoric and some actions are shifting towards a more hydrosocial approach, much more activity to 'reconnect with water' is needed if this vision is to become a reality. 'Hydrosocial water management' was identified as differing from the currently dominant technocratic model because of its engagement with stakeholders and a humble perspective that accepted that there are many different valid perspectives on water management.

This book has explored the potential contours of this new hydrosocial approach, through an examination of urban water management and its changing form. In so doing, it has ranged between centuries and continents, focusing in each chapter on different parts of the water cycle, and drawing particularly on an emerging body of interpretive social science about water. In this conclusion, I bring this material together, articulating the promises of a hydrosocial water management, assessing the ways it has already been achieved, and identifying routes forward to hasten its development. Finally, while much of the chapter is focused on new perspectives on water (and hence on how this book has sought to shift understandings in the engineering-related field of water management), in the last section of the Conclusion I reflect on the contributions of the book to social science.

Hydrosocial water management

Emergence

The recent emergence of some hydrosocial approaches to water management needs to be understood in the long span of changing human relationships with water, linked to the development of cities. Urbanisation created significant water challenges for society. This is because the concentration of human activities in one location overwhelms the cleaning and recycling abilities of the natural environment. Specifically, waste water from human sewage, animal waste and manufacturing processes caused deterioration in the quality of water in streams and rivers, creating problems with clean water supply (Chapter 3), as well as public health and the aesthetic qualities of rivers and their surrounding cities (Chapters 6 and 9). Also, the increase in impermeable surfaces exacerbated risks of flooding and further pollution, with the potential for human disturbance and distress from losses of increasingly valuable urban buildings (Chapter 7). In terms of the social processes involved, many of these developments were interlinked. For example, as described in Chapters 4 and 6, the provision and marketing of piped water drove the development and spread of toilets, which then encouraged waste water to be placed in sewers, which exacerbated problems of water contamination.

Each of the water challenges stimulated a variety of potential solutions. Generally, the solutions that were implemented involved significant technological change. As we have seen in Chapter 3, water supply was addressed through accessing sources further away from the city, as well as through increasingly sophisticated treatment processes. Addressing water quality issues involved the development of a sewage system, as discussed in Chapter 6, but also the abandonment of urban watercourses to become drainage ditches or sewers. Flood management involved the development of flood defences (FDs) as outlined in Chapter 7. Some solutions – those we still know about – such as the provision of piped water, spread and became the hegemonic means of ensuring humans can access water. 'Technocratic water management' was the term given in Chapter 1 to describe this collection of technical solutions to water problems resulting from urbanisation.

In Chapter 1, and again in subsequent chapters, I argued that this technocratic water management has come under critique, particularly in the last 30 years. Conceptually, one important critique was that these processes were not (and in many cases still are not) carried out in an interconnected manner. As implied by the nirvana concept, 'integrated water resource management' (IWRM), despite the conceptual linkages made in the hydrological cycle, management of water occurred in sectorally isolated areas of 'flood defence', 'water supply' and 'river management'. Another important critique, particularly championed through the concept of ecosystem services, is that water systems deliver many benefits and have some costs beyond their immediate institutional context through their maintenance of a flood resilient, biodiverse and aesthetically pleasing environment. In this respect, the environmental, social and aesthetics impacts and benefits that potentially result from the water system are highlighted, and it is argued that they should play a more significant role in water decision-making. A third conceptual development was that of 'adaptive water management', particularly stressing the need for more learning and flexibility in the institutional systems for water management, and the need to move on from single one-size-fits-all solutions.

In terms of the negative effects of technocratic water management on the ground, as recounted in Chapter 3, in the area of water supply there has been increasing concern about unsustainable abstraction and related damage to ecosystems. This has driven some water-aware cities to focus on alternative measures, including demand management, recycling of water and the potential for developing more decentralised components of the water system. Likewise, as discussed in Chapters 7, 8 and 9, in the field of flood management there has been a sustained critique of dominant FD approaches, and a new flood risk management (FRM) paradigm has placed more emphasis on working with nature and with the public to reduce the probability of flooding but also to manage its potentially damaging effects. The management of water quality in our rivers and streams is perhaps the least 'advanced' of these fields of change; as discussed in Chapter 6 there is a recognition that the former emphasis on point sources of pollution is not sufficient to improve the quality of rivers and streams, and that diffuse pollution from agriculture and urban areas also needs to be considered. Translating this criticism of technocratic ideas into new practices has been difficult: for example, there has been little change in the practice of draining roads and streets directly to watercourses. Nevertheless, as argued in Chapter 9, attention to the quality of the landscape around water in our urban areas is gaining attention, and is increasingly recognised as requiring public engagement as well as technical restoration of habitats. In each of these different watery fields, then, there are some critiques of technocratic approaches, and attempts, albeit halting, to translate these into practice. Hence, to some extent, change is underway.

For Linton (2014) and for me, this emerging type of water management differs from traditional technocratic approaches not because it looks at different parts of the water cycle together (which it does not, or at least, not always), nor because it pays attention to the environment (which it generally does), but because it focuses on working with people. Indeed, 'hydrosocial water management' requires a complex interactive human governance process to prioritise and deliver water management. Specifically, the goals of water management are not simply technical, but balance different potential technical, ecological and other goals linked to different social expectations. Likewise, delivery of water management is not just a question of organising the appropriate physical intervention, but balancing this against (and/or combining it with) potential social interventions, including attempts to renegotiate expectations and service levels. Hydrosocial water management, then, is this collection of activities that seeks to work with the public and other stakeholders, to actively manage water, and hence to view water management as part of a social system rather than as an isolated technical process. Moreover, as argued in Chapters 1, 7, 8 and 9, these 'engagement' processes need to be conducted in a spirit of humility. In other words, the scientific and technical knowledge of, and expectations about, the water system need to be viewed as one form of knowledge that sits alongside other unexpected 'lay' ways of understanding and knowing the system.

Hydrosocial water management can be seen as an emerging set of ideas about how water should be managed. As will be apparent from the discussions that follow, while we can say something about how this approach works, there is much more to be tried, discussed and delivered. Moreover, while some aspects of this approach are beginning to be implemented, the concept requires further development, and it is far from universally known, let alone applied.

Should the hydrosocial approach to water management be viewed as yet another potential 'nirvana concept', claiming the supremacy of its ideas but in practice competing with other nirvana models of water management for the attention of practitioners and academics? The answer is no. As I argue below, hydrosocial water management's sound foundation based on empirical understandings about how policies and practices work and change means that the concept is both more realistic and more likely to be implemented than these other nirvana concepts. First, however, it is important to discuss more fully what hydrosocial water management means, and how its vision differs from the current technocratic model, as well as from those other nirvanas.

Characteristics

So, what are the central characteristics of this hydrosocial model of water management? As we have seen, the hydrosocial model is counter-defined against the technocratic model because of its 'social' characteristics. This suggests that those delivering and regulating water services are doing so in a way that explicitly interacts with (that is, reflects, parallels and influences) changes in social understandings and practices. The water systems are therefore conceived of as truly 'socio-technical' and/or 'socio-ecological'.

Some specific implications of these social characteristics are:

- *Water services are a matter of public debate and local experimentation.* Big choices about how we want our water systems to run are a matter of public debate. Local experimentation – trying out new ways of doing things and learning from what went well or less well – is an important part of this process. Water service providers (whether private or public, whether providing water supply, drainage, or sewerage services) and their regulators are likely to play a leading role in stimulating and organising public debate about significant water-related choices that need to be made, especially concerning investments and/or changes to service levels and expectations. Water utilities, regulators and civic society organisations all take a role in stimulating and driving these changes.
- *Water utilities and regulators undertake social innovation.* Negotiating and supporting changes to public behaviour and expectations is a normal role for water utilities, and is recognised as such. Potential changes to social expectations and practices are considered alongside technical developments and investments as means of addressing the challenges faced in water management. Likewise, the socially innovative aspects of technical innovations are fully recognised, and hence prepared for and managed. Ideas to 'open up' potentially different ways of managing water systems may emerge from stakeholders within the system, by learning lessons from historical water systems, or through being inspired by

practices in other localities. Designing and implementing schemes to provide changed institutions for water management may involve existing water service providers directly, or it may be best achieved through work with partners including those in the private sector (e.g. plumbers) or civic society organisations. Substantial developments in the nature and form of water services involve long-term interactions between a water utility and its publics, with years of attention to developing trust and highlighting choices so changes are fully understood by both sides as part of a developing 'hydrosocial contract' (Wong and Brown, 2008; Murphy, 2014).

- *Water utilities use and develop social knowledge in the design and evaluation of innovations.* Water utilities need specialists in understanding, interpreting and generating social knowledge if they are to play a part in public debates and negotiate and appropriately design and learn from both social and technical innovation. The information they need might concern public practices, but equally it might concern motivations, attitudes, or attachments; it might concern their general customers, a specific sub-set thereof, or it might concern other groups of people such as their staff. They need social science-trained staff internally in order to generate and use this knowledge. The utilities additionally need to be empowered to draw on and learn from external academic social scientists who, like their technical counterparts, play a role in guiding and supporting the organisations in designing, developing and evaluating innovative practice.
- *Water services are tailored to specific localities.* A diversity of local physical and institutional systems will be used for delivering water services. Though the traditional centralised water system delivering a universal service is likely to continue to exist in places, it will increasingly be augmented by: (1) more diverse physical systems for water storage, water supply and drainage, including, for example, household rainwater harvesting, dual pipe systems, sustainable drainage features and community-level water recycling and (2) more diverse and decentralised or partially decentralised institutional systems for owning and managing decentralised elements of the water system – for example, central requests for household water tanks to be emptied as a flood precaution in anticipation of extreme wet weather. The diversity results from different underlying hydrological conditions, physical systems and cultural norms, and also because opportunities for investments and experiments in the systems will arise ac different times.
- *Municipalities identify, pursue and support the realisation of co-benefits between different services, including water.* The significant potential for water system investments to augment other environmental and social priorities (and for water systems themselves to be augmented through other investments) creates a role for municipalities as well as local civic society organisations in identifying, developing and implementing multifunctional solutions that address a variety of local issues in innovative ways. Both funding models and allocated responsibilities need to provide the freedom to exercise these roles.

- *People know about their local water systems.* Compared to the contemporary situation, the public is better informed about water systems in general and in their locality specifically. This knowledge develops because:
    - the public have contact with a variety of water systems as they move between different neighbourhoods and towns in their daily life
    - water service providers make the specific local peculiarities of their systems more evident to the public than at present – for example, providing opportunities for school children to access and to assess figures about water locations, quantities and qualities near their school
    - a context in which there is significant media and civic debate about water-related choices augments public understandings.

There are some respects in which what is described above is similar to the sort of water management described in Chapter 1 under the 'nirvana' concepts of 'integrated water resource management', 'ecosystem services' and 'adaptive management'. In particular, the sorts of physical systems envisaged coincide with (and have been partly inspired by) those anticipated by adaptive water management and water-sensitive urban design. The big differences between the conception of a good water system envisaged by the hydrosocial model and what has already been explored through these other concepts are:

1. a greater emphasis on the social and political context in which water management takes place and, in particular, a greater faith in the possibility of achieving changes in social expectations in a specific locality rather than just having to meet universal expectations imported from elsewhere; likewise, an acceptance of the possibility of mobilising publics to achieve changes in practices;
2. an acceptance of the relevance and importance of information about social expectations and aspirations, alongside ecological and engineering science in informing the decisions made. This information might take the form of the outputs of engagement or feedback processes (whether carried out by the water organisation themselves or others with whom they are working in partnership) or it might involve more in-depth interpretive social science research findings;
3. along with adaptive water management, an emphasis on experimentation and learning in order to develop and deliver locally appropriate tailored solutions to specific local water challenges.

Current practice

To what extent is the hydrosocial model already being applied in managing water in the cities of the industrialised world? In terms of current uptake it is certainly the case that many institutional processes are increasingly recognising the need to find ways for the public and other stakeholders to participate in exploring decisions about water management. There are also increasing attempts to mobilise specific groups of the public, or other stakeholders, to try to change their actions to address water-related challenges. Some of these participation and mobilisation processes draw on principles of good participation such as those introduced in Chapter 1 and discussed again in Chapters 8 and 9. It is also the case that, over the past two decades, research on difficult water management issues have frequently involved social scientists as well as physical and engineering scientists. In this respect, there is a growing body of social science on water management, much of which comes from a broadly interpretive standpoint. There are pockets of practice where this social science is being recognised and used within water management practice. This book has illustrated some examples of what might be considered good hydrosocial water management, as shown in Table 10.1.

In contrast to the examples in Table 10.1, however, many institutional water processes that include social and political decisions are framed as purely technical and financial within the current decision-making process. Likewise, most water management challenges are considered in terms of the ease of addressing them technically first; only subsequently are any potentially social processes considered. And while mobilisation and participation initiatives are carried out, many are designed and run by engineers who have limited experience or knowledge of public engagement; evaluation is limited and focused on internal arguments and

TABLE 10.1 Examples of 'good' hydrosocial water management from preceding chapters

(Chapter) Example	*Why it exemplifies hydrosocial water management*	*How it might have been even more hydrosocial*
(5) Research on water use practices with potential to inform water-saving promotion	Seeks to understand water use processes from the user's point of view before seeking to influence them	Active exploration about how the learning from this work can be applied in different localities
(6) Waiheke's decentralised water supply and treatment systems	Illustrates diversity of water systems; a peculiar local way of managing water is preferred by local people	More active discussion about water quality in rain tanks and more monitoring of impact on watercourses
(6 and 9) Friends of Bradford Beck actions on water quality and the role of the river in the city	Community group-driven mobilisation of institutions and the public to improve local river	Earlier and more enthusiastic participation from water management organisations
(7) Integrated urban drainage work in West Garforth	Researchers, practitioners and the public work together to address an apparently insurmountable local flooding issue	Funding streams could have been made available that support many small sustainable urban drainage systems measures
(7) Research on flood experiences in Hull	Research seeks to understand people's experiences and to inform policy in Hull and beyond	Such research could be structured in to any process of emergency response in order to support learning

water-related goals and there is no collation of learning between water organisations in order to build a science of water engagement. Finally, and most importantly, while there is a profession of scientists who are trained and accustomed to using engineering science and ecology working within these water organisations, they are seldom supported by individuals who have undergone social science training. Consequently, utilities are in a weak position, unable to engage with social science that can improve their understandings, evaluations or practice.

It is important to stress what is potentially being lost here. Socially innovative practice, and evaluation thereof, has occurred in a myriad of other sectors (e.g. social services, energy management) and led to great expertise on collating and understanding social information, and designing, supporting and evaluating social innovation to tackle apparently technical management challenges. As long as interpretive social science is dismissed as anecdotal and peripheral, water management it is cut off from this potentially highly enlightening well of skills and knowledge. Learning is required on all sides, however, not just by water managers. Much existing social science on water is developed and pursued from an external critical perspective, with little potential to impact on water management practice. Interpretive social scientists too need to change, to grasp the opportunities to work with open water managers from a variety of institutions in order to explore together when and how society can benefit through applying social science to water problems.

In general, it is fair to say that current water management remains predominantly technocratic. Table 10.2 identifies some example narratives from this book that illustrate the ongoing importance of the technocratic approach to water management.

Hydrosocial water management and the nirvana concepts provide alternative sets of ideas about how contemporary water management should change and develop. The nirvana concepts, however, obscure a real choice to be made between technocratic and hydrosocial routes forward. In some respects, the nirvana concepts allow space for a hydrosocial approach to emerge: for example, an ecosystems services approach that seeks to identify benefits and costs of different options to enable a political decision, or an adaptive approach that collates stakeholders' views together to reflect on learning from a real-world water management experiment. But many of their suggested solutions remain predominantly technocratic – for example, both the 'integration' of water management choices through multi-criteria optimisation and the monetisation of ecosystem services in order to make decisions about co-benefits and trade-offs can be seen as expert-led technocratic approaches. The difficulty with the nirvana concepts is that, because they leave this choice unclear, the damaging and destructive impacts of technocratic water management have the potential to continue, albeit presented within the rosy glow of one or other nirvana concept.

In the context of a world of water management institutions dominated by those with a technical education, hydrosocial practices and concepts might be regarded as having made surprising progress and, at least nominally, to be recognised as relevant to decision-making. But we can do better. Social innovation via participation and

TABLE 10.2 Examples of technocratic water management

(Chapter) Example	*Why it exemplifies technocratic water management*
(2) The Falkenmark index and other measures of water scarcity	The Falkenmark index and other measures of water scarcity are based on global statistics and averages. While they have a significant communicative value, they pressure a relationship between water use and well being which is questionable, and hence obscure local culturally specific solutions that might address scarcity issues
(3) Fossil-fuel powered desalination as a means to address water scarcity	While providing an increase in the range of options available to a water-scarce city, desalination plants entail high financial and environmental costs both for their construction and their running
(5) Behaviour-change approaches to water conservation	These blame the user for high water use when arguably the conditions for high water use are created by the system that provides it free at the point of use
(7) In West Garforth funding mechanisms made it hard to address local flooding issues	While prioritisation of investment funding is inevitable, the investment system favoured large schemes saving many houses and provided limited support to address surface water flooding, which tends to affect fewer houses more chronically
(9) Malin Bridge vegetation clearance for flood alleviation	Although consultation was carried out, it was a limited effort and did not really allow room to change the plan

mobilisation should not be just an occasional experiment, carried out by engineers when all of the technical options have run out. Instead, participation should enable water utilities to be working in cooperation with their publics; opportunities for social mobilisation and the renegotiation of service levels should be considered alongside technical solutions in relation to each dilemma faced. Likewise, social science does not need to be smuggled in under the Trojan horse of technical research projects; it should be invited in to support and evaluate processes of social and technical innovation. The ongoing challenge is to transfer the previous openness to aspects of a hydrosocial approach to achieve such a step change in discussions, practices, science and employment. Before this can be done, it is important to consider the risks associated with a pseudo-hydrosocial approach.

Risks

Despite the preceding dominance of the technocratic approach, this book has taken a largely optimistic and positive view to argue that the hydrosocial approach now has a real foothold and that there are genuine prospects to embed a more socially informed form of water management (of which more below). But within the family of interpretive social science approaches, writers from political ecology might be expected to view these claims with some scepticism. And, indeed, the history of nirvana concepts highlights how, unless conceptual clarity is achieved, terms can be misappropriated and used to operationalise and develop anti-progressive practices. What, then, are the aspects of hydrosocial management that raise risks and could be used to smuggle in technocratic or non-progressive elements? What actions need to be taken to support the hydrosocial approach, while also guarding against its misappropriation?

Two specific risks can be identified. First, from a political ecology perspective, mobilisation might be viewed as a route through which utilities, whether privately or municipally owned, can avoid responsibilities that were previously delivered by them, or by the state. This critique is part of a general concern that right-wing forces are pushing back the boundaries of state intervention, and that an equality of citizenship that was previously achieved will be reduced through new neo-liberal politics. The second critique relates to the hydrosocial approach's advocacy of new and differentiated service levels suitable to different geographies and cultures. By breaking down universal service levels, a political ecologist might argue, a marketised system of differentiated services will be smuggled in and the poorest will suffer as they will be unable to pay for basic quality water services.

This book has illustrated examples of mobilisation being used by water utilities to encourage water saving and to help make members of the public into managers of water services – for example, through managing water tanks for water supply and/or for mitigating heavy rainfall. It has also shown how civic society groups can use mobilisation to perform 'public' services such as litter picks, water monitoring and policy coordination and lobbying. Effectively, the critique of mobilisation argues that these processes involve users being asked to take responsibility for water services, and to bear the blame when things go wrong. It also suggests that users who cannot or will not be mobilised will suffer differentiated service levels and poorer water services.

The first defence against this critique is to highlight that asking the public to take action is not in itself a bad thing. Water services are both technical and social; social behaviours influence the services and if, by different behaviours, we (the public) can improve the services for everyone we should at least be aware of it. Indeed, by not asking us to take some responsibility the system is effectively disempowering and disbelieving in us. The critical point comes, however, if utilities start blaming users when there are problems with services. Zoe Sofoulis and her colleagues have already suggested that this happens when requested water savings were not delivered (2005), and one can envisage the same process of user blaming would be possible in relation to use of waste water services, or in relation to waste water quality. The key point is that there is nothing wrong with asking the user, but there is something wrong with blaming the user. In this respect, responsibility has to remain with the utility. As illustrated in Australian examples from Chapter 5, respectfully framed flexible requests to users can be far more powerful than inflexible commands. A central need here is for understanding, both of how the water service is operating and of the logic and choices the water service users are making. As illustrated in Chapter 5, interpretive social science can be key to exploring water user perspective and in designing campaigns and messages. We also need to ask how the water system became so dependent on the behaviour of users without them understanding. An adaptive hydrosocial approach would see any such potential 'blaming' situation as a learning opportunity that should prompt the utility to revisit expectations, communications and relationships.

Moving on to the second critique, a central concern of hydrosocial water management is that we reject the idea that 'one-size-fits-all' to expect different types of wates systems in different locations. This means that service levels concerning, for example, minimum pressures, expectations about the colour of water, ideas about water quality both in the tap and stream and risk statements about levels of protection from flooding and drought are all potentially brought into question. The political ecologists are likely to be concerned that those users who are not in a strong a position to negotiate will suffer from a deterioration in water service levels as they will be unable to pay and also unable to be mobilised.

Of course, the idea of universal service levels is an illusion. What we actually have are minimum service levels, which sometimes differ between countries. Meanwhile, the service levels achieved always vary within countries. Service levels are important as they offer benchmarks and expectations about what utilities will deliver. For example, in the UK there are service-level expectations about the proportion of properties at risk of different types of flooding, the number of water supply quality failures and the quality expected in rivers and streams. However, the service levels also provide a straightjacket, dictating the way water services develop in one locality, even when hydrogeology and culture might suggest other ways of meeting water service needs. For example, a rule that every house must be connected to water supply system would mean that the potential for using household rain tanks is seldom considered.

What this critique demonstrates is a genuine tension between, on the one hand, the need for social equity and fairness in the provision of water services and, on the other hand, the need for innovation, including the potential understandings of desirable service levels to shift between places and times. It is not simple to identify the appropriate balance between these two factors, and certainly it is not possible to dictate a rule of thumb for future choices. What can, perhaps, be suggested is a process through which the nature of this balance is negotiated. First, service levels are only likely to be called into question when there are significant water challenges that are faced. As discussed previously, these are societal challenges that should not be viewed merely as the problem of the water utility or state; in this respect public participation, awareness and discussion of challenge is key. Second, in exploring ways that water challenges can potentially be met, examples drawn from alternative times and alternative cultures can be instructive, providing 'left field' ideas about how problems can be addressed. Third, a combination of playfulness and science may be needed to explore whether some of these less conventional ideas can be applied in a contemporary industrialised country context; for example, can ideas about different levels of water services be tried out at festivals, campsites, or some other 'temporary' venues where learning can occur (Browne, 2015)? Obtaining informed consent from participants, and then collating their views and reflections of the experiments will be key to ensuring that learning occurs. An important part in this process needs to be taken by regulators. Rather than just being charged with exclusively health and safety or environmental concerns, regulators need to be expected to work together to help water utilities address the long-term sustainability of their services. This requires them to explore and investigate different models for service provision, including those that might involve radical change and social innovation. It also requires them to examine radical shifts in regulatory models in order to facilitate changes and for experiments to occur. Rather than detailed specifications about service levels, discussion is needed about the universal principles that should act as appropriate yardsticks for water services; for example, we might expect that households have 'adequate (quantity and quality of) water', and that 'sufficient protection for biodiversity' is provided.

Finally, the political context for these risks needs to be noted. Writers from the political left, such as political ecologists, are often concerned about what they see as a dilution of what has traditionally been done by the state. Their position plays into a version of politics that sees big state and significant state interference in the market as a left-wing expectation, whereas minimal state involvement or interference is the preferred approach of the right; in cultural theory terms it is about hierarchical or egalitarian politics rather than the individualist or fatalist politics of the right. This critical perspective would view developments in the water field in this light, with any changes or reductions in the field of water management perceived as the retreat of the state and the further delegation of responsibilities back to citizens and towards individualist and fatalist ways of organising. Left-wing and anarchist defenders of hydrosocial changes, however, would regard changes towards hydrosocial water management as a move from hierarchical control towards more egalitarian politics. The question as to how the hydrosocial approach can be protected then becomes one of how individualist or fatalist politics of water management can be avoided and egalitarian forms of organising supported.

Contributions of interpretive social science to the hydrosocial approach

Interpretive social science has been presented as the bedrock of the hydrosocial approach. This section will explore four contributions to hydrosocial water management in turn: first, its stance on knowledge; second, the opportunities posed by its empirical findings to open up debates; third, its understanding of social processes; and, fourth, the light it throws on prospects for the hydrosocial approach itself.

Stance on knowledge

As discussed in Chapter 1, interpretive social science is understood broadly to include a range of post-positivist research and theorising, some of which might be explicitly positioned as 'interpretive' but much of which is not, positioned instead as 'political ecology', 'science and technology studies' or just as 'constructivist'.

In the context of an interdisciplinary readership, the complex debates that distinguish one variation of this approach from another are extraneous detail. The common features are that all such approaches reject the assumptions of positivism. In particular, interpretive approaches make no assumption that changes in the human condition represent 'progress'. Moreover, it is recognised that there are always different valid ways of studying and understanding the world. Processes of meaning-making are at the centre of interpretive enquiry; this necessarily means the approach is open to different interpretations. Importantly, however, knowledge needs to stand up to scrutiny. As discussed in Chapter 1, a test of robustness in interpretive science is the extent to which the reader is convinced that the researcher is transparent and honest about their positionality and the way that this influenced their interpretation of events. In this respect, as a reader of interpretive social science you are invited to ask: 'does this account ring true as the writer's honest and reflective account of what he/she studied?' You might assess honesty in terms of whether the writer provided evidence, so that you feel like they have fully understood and engaged with the subject. In terms of whether the writer is 'reflective', you would be likely to consider what the writer has said about their own positionality and whether you believe that they are fully transparent about its influence on the interpretation presented. Ultimately, readers are likely to judge a piece of work robust if it is practically useful and accords with their lived-reality and problem perceptions.

Interpretive social science's stance on knowledge is, perhaps, the most important but also the most challenging contribution to defining hydrosocial water management. As with many technical fields, there can be a scientific conservatism to technically trained water managers who are accustomed to traditional positivist scientific ways of making knowledge, based on the need to generalise in pursuit of universal truths. There is a tendency to dismiss knowledge that is seen as anecdotal or personal experience, in particular, if it challenges or undermines existing understandings or practices. The effect of this positivist stance on knowledge can mean that public participation and other hydrosocial water management techniques are regarded as threats rather than opportunities.

If water managers take on an interpretive approach to knowledge then they will see lay knowledge alongside the knowledge of their fellow water managers and those working in allied fields (like energy management) as valid and useful sources that can inform their practice. Crucially, they will be aware of their own 'positionality' – that is, the way that their personal background, training and experience is influencing the assumptions and understandings they bring to their work.

Opening up debates

Interpretive social science's second contribution to hydrosocial water management is to provide empirical evidence that can help to open up debates and ideas about how water might be managed in the future. The process of opening up different possible futures occurs through narratives about either what has happened in the past or about what is happening now in other locations. For example, Table 10.3 identifies some of the popular myths about water that the accounts in this book have sought to dispel. The extent to which it is likely that people want to act on these narratives will vary.

The list in Table 10.3 is far from exclusive in terms of the myths that might be busted through evidence, but illustrates the importance of history and geography in opening our eyes to different possible ways of organising things. The issues have the potential to help stimulate debates about whether and how our water systems might change today and into the future. Not every narrative will have resonance and lead to change, but some may. This is particularly important with an infrastructure service like water, where there is limited folk memory of how people managed before the technical system existed (for illustration, contrast with the internet, for which another 50 years will need to pass before those who remember pre-internet days are no longer around).

It should be stressed that it is not just interpretive social science that dispels myths about water. Other forms of research – for example, more positivist forms of social science such as Hoekstra's work on virtual water reported in Chapter 2 – make

TABLE 10.3 Water myths dispelled

Water myth	*(Chapter) Dispelled because*
Water supply emerged in order to address the public health problems of the poor	(3 and 6) The provision of water supply was largely driven by commercial concerns
Washing is a necessity to human life	(4) It is clear that people lived reasonably satisfactory lives without washing
Human development involves becoming progressively cleaner	(4) In Europe, at least, this is palpably not the case, with two centuries when personal cleanliness (as we perceive it today) was not required, though clean clothes were!
Sewage systems were developed because they would stop the water-borne transmission of disease	(6) Sewage systems were developed in order to address smells that were perceived as a vector for disease, and hence to relieve the poor's call on the state
Famous UK public health reformers fought together to achieve a better water system	(6) Chadwick, Snow and Bazalgette may be perceived in retrospect as together pushing for a better sewage system, but, in practice, Snow was ignored by a health establishment, within which Chadwick and Bazalgette fought on opposite sides of heated debates over the nature of sewers
Human waste must be transported and treated as a liquid	(6) Up until the early nineteenth century, human waste was treated as a solid and used for fertiliser (something that is challenging today if it is mixed with heavy metals from road drainage before it is treated)
It is not possible for urban areas in industrialised countries to function without centralised water supply and drainage	(6) As demonstrated from the example of Waiheke, a suburban area can function quite satisfactorily without centralised water and drainage

important contributions to our understandings about the high proportion of water that is 'consumed' via food, and the role of virtual water imports and exports in driving environmental destruction in some locations. The use of engagement techniques also has the potential to open systems up to different ways of doing things that had not previously been considered. But interpretive social science provides narratives about water services in alternative cultures that may make other ways of managing water imaginable.

Understanding social processes

The third contribution of interpretive social science is to our understanding of social processes. The standard interpretive retort to the criticism from positivists that interpretive anecdotes are generalisable is that we generalise to theory, not to numbers. Theory is, effectively, a shorthand 'rule of thumb' about how to understand a social situation, derived from one or several empirical instances but with an implicit claim that the same understandings may be useful in other contexts.

There are several important examples of the use of theory in this book.

- Chapter 2 developed the 'hydrosocial governance cycle' (Figure 2.5). This moves beyond the immediate human impacts on the water cycle provided by the traditional water cycle (or variations thereof) to conceptualise how governance and people interact with the physical environment. Specifically, it conceptualises waters as part of the physical environment but does not separate the natural and engineered aspects of the latter, which, as Chapters 3, 6 and 7 have illustrated, is not possible to do. Moreover, it highlights the interactions between organisations and concepts in the processes of governance. It is a relatively simple diagram, but it offers a framework of pointers through which any water management situation can be analysed.
- In Chapter 3, I introduced ideas about different aspects of integrated governance. Focusing on the topic of water supply, I discussed how at different times and in different places water supply operations and regulation had been integrated (1) at different scales and (2) with different water and non-water functions. One form of integration often comes at the expense of another. For example, geographical integration of water management in the UK that came with the creation of the regional water boards also came with water being cut off from issues of local aesthetic interest and well being associated with local authority control. The rest of the book has revealed similar variations in relation to the governance of water quality and of FRM. In developing and using theory about integrated governance, it becomes clear that there is always a case for a new and different form of integration: for working more closely with another industrial sector, or for looking across boundaries in a catchment, for example. But integration of everything is not possible. Hence, we are faced with a constant cycling of different forms of integration and different emphases in terms of who/what work together. Particularly in the context of calls for

more integrated working across infrastructure sectors, this insight highlights how the question is not 'whether we integrate', but rather 'what we integrate' and what we leave as separate domains.

- Chapter 8 extended the discussion of governance to explore how a framework of 'governance cultures' can support policy comparisons, for example, between different types of FRM. It was explained that cultures vary in the way they understand the external risks posed by nature, and also in the way that they view human nature. It was suggested that while ownership is often used as a proxy for understanding governance culture, empirical understanding of what actions are undertaken and what communications are made enables governance culture to be understood more fully. The change from FD to FRM was understood as a shift from an explicitly hierarchical culture towards one that was explicitly more egalitarian. Underlying such explicit aspects of governance are the parts of FRM that are deemed too expensive or not important enough for investment: implicitly left to people's buying power to provide insurance or to build resilience for their own households, and hence an individualistic governance style. Though applied in Chapter 8 to FRM, it was argued that governance cultures offer an important and useful framework for understanding different approaches to water management. In Chapter 9, and below, I have argued that governance culture can help us to understand some of the differences between approaches emphasising mobilisation and participation. Elsewhere I have used the same approach to look at mobilisation initiatives relating to water and energy efficiency promotion (Sharp *et al.*, 2015). Overall, the governance culture framework offers a general means of analysing how collective goals are achieved that can, therefore, support comparisons between policy areas (e.g. flooding and water supply), localities and initiatives. In terms of supporting practice, the framework could help a reflective practitioner consider the implicit culture of initiatives or activities that they have undertaken and hence to make informed decisions about how to develop future initiatives.
- Chapters 4 and 5 discussed and exemplified practice theory, suggesting that, rather than highlighting behaviours – which focuses exclusively on the consumer as the potential agent of change – instead, it is useful to discuss water use practices. The fundamentals underlying any practice are its different components – the material 'stuff' used, the procedures undertaken and the inherent purpose or meaning. An important insight is that these inputs to practices are controlled and managed by many different external bodies – for example, water and energy companies, supermarkets and manufacturers of toiletries – hence, the consumer is not the only potential agent of change. If the practice is problematic for the environment or in some other way, the practices approach invites the analyst to ask which of the factors can be shifted to generate the least harmful effects. Applying practice theory enables new and different stories to be told about the past. For example, we can see the process of cleaning ourselves as one where practices have shifted dramatically through time, with the

period 1500–1750 identified as one in which it was clean linen rather than clean bodies that was valued. Practice theory also provides a powerful analytical tool to understanding current water use as a potential guide to new and more focused processes of mobilisation, as indicated by the work of Alison Browne and her colleagues (Browne *et al.*, 2013b; Pullinger *et al.*, 2013). Overall, practice theory seems to offer an analytical tool to look beyond broad generalisations – for example, about the average water use per household – and hence to understand important mechanisms of change in our society.

- Chapters 5, 8 and 9 addressed issues about engagement. Engagement concerns the way that water organisations work with others, whether these are professional stakeholders or the wider general public. As noted in Chapter 1 and reiterated in Chapter 9, the theoretical insight is that engagement can involve either or both 'mobilisation' (seeking to influence other's actions to achieve a water-related end) or 'participation' (collating other's perspectives to influence a water-related goal), or both. Though much discussed in other fields, water scholarship lacks critical discussion of engagement strategies including both participation and mobilisation. In a context in which engagement is increasingly seen as an alternative to technocratic management, this is clearly a significant omission.
- In terms of mobilisation, Chapter 5 highlighted a contrast between traditional behaviouralist approaches to addressing water issues involving the consumer and more recent scholarship on practices, already discussed above. Using the governance framework presented in Chapter 8, we can now understand the behaviouralist approaches as linked to an individualist governance culture, while the practices approach is more egalitarian. The theoretical insights is that mobilisation can be framed in terms of individual benefits or common benefits. In Chapters 7 and 9, some examples of mobilisation were discussed: flood victims in West Garforth were mobilised to take action on their own flood resilience; likewise residents in Bradford are mobilised to clear litter from their local watercourse. Notably, these are both examples where mobilisation occurs alongside participation (and hence might be seen as exemplifying a more egalitarian governance culture, associated with achieving common benefits). Overall, we can understand mobilisation as an increasingly important part of water management and a crucial element of the hydrosocial approach. Much work remains to be done exemplifying when, with whom and through which mechanisms mobilisation is best carried out.
- In terms of participation, Chapter 9 explored some theory of participatory processes, understood to be associated with the left-hand side of Table 8.1 and hence with hierarchical or egalitarian governance cultures. Mark Reed's axioms were drawn upon to highlight some of the conditions for high quality participation (Reed, 2008). Four examples of cases where participation might have occurred were presented in Chapter 9; additional examples can be seen in the West Garforth and Hull cases of flood management, discussed in Chapters 7 and 8. The cases showed that while participation was lacking (Dearne restoration)

or tokenist (Malin Bridge) in some examples, in others the public had clear input to policy-making, whether this was invited (West Garforth and Hull) or claimed (Bradford Beck and Five Weirs Walk). They demonstrated that technical expertise was spread through society, and also that a close emotional engagement with the landscape was experienced by professionals as well as by the general public. In this respect, the discussion highlighted how the distinction between 'stakeholder participation' and 'public participation' that is frequently perpetuated in the literature may be unhelpful. In practice, 'the public' are a diverse set of stakeholders with a variety of perspectives and knowledge relevant to water decisions. Overall, an understanding of participation as a form of governance and of the conditions that enable it to be effective provides clear and useful guidance that can support both researchers and practitioners in understanding and designing future engagement initiatives.

In terms of hydrosocial water management, my general claim is that these theories enable a more informed and nuanced understanding of a particular situation, and hence can support future analysis and better policy design.

Supporting social innovation

Interpretive social science supports social innovation when social science theory and understandings inform new and experimental practices. Examples in this book of where this has occurred include the research in West Garforth (Defra, 2008) that supported participation and action for flood mitigation in West Garforth, and the work by Woelfle-Erskine (2015) that informed innovation around water resources in California. Other examples drawn from the literature where interpretive social science has informed social innovation include the work by Lane and his colleagues in Sussex and East Yorkshire (Lane *et al.*, 2011), my own work focused on integrating new surface water management practices in Dyr Cymru Welsh Water (Westling *et al.*, 2014) and potentially the work by Browne and her colleagues that has informed Defra's thinking about water management (Browne *et al.*, 2013a and b). However, as discussed above, as a whole there has been relatively little use of interpretive social science to support changes in water practices.

Prospects for the hydrosocial approach

The final area in which interpretive social science contributes to hydrosocial water management is in its overall understanding of policy processes. Through this understanding, interpretive social science can provide an assessment of the prospects of the hydrosocial approach and advice to those of us who may wish to support its development.

Interpretive social science is well established in the field of policy analysis (Hajer, 1995; Hajer and Wagenaar, 2003). This field highlights how we can understand governance in terms of the contest between the ideas and meanings given to the world by different stakeholders. Stakeholders, either singly or in groups are likely to be associated with: (1) specific ideas and narratives, (2) formal institutions, such as academic disciplines, companies or regulators and (3) a set of practices. Input from the sociology of science and technology has stressed that material objects (4) can also perpetuate the notions of a particular sets of policy ideas and practices.

This book has already applied this interpretive policy understanding to the analysis of the field of water governance in general, and the governance of water supply, water quality and FRM in particular. It has argued that, broadly, we can understand that there is a hegemonic technocratic approach to water management that is deeply embedded in many water-related expectations (e.g. the expectation that a road will be drained), practices (e.g. flushing the loo), institutions (e.g. the strong association between water management and civil engineering) and structures (e.g. the sewage infrastructure perpetuates a lack of engagement with human waste). It has highlighted how technocratic water management is now under critique. It has suggested that hydrosocial water management provides a sustained and coherent set of ideas and potential practices through which such technocratic approaches can be challenged.

Having reconceptualised the field of water management, what does interpretive social science have to tell hydrosocial water management about the process of policy change? In other words, what factors have to be borne in mind by those who wish to support a shift towards a hydrosocial approach in order to help bring these changes about? In the final part of this subsection I identify some general principles that can be drawn from other work on policy-making. These are translated into some specific challenges for hydrosocial management discussed in the next section.

Three key lessons can be identified.

First, struggles between different sets of policy ideas (and their associated institutions, terminologies, objects and practices) are ongoing, and played out continuously in many decisions, ranging from small local decisions over how a park is redeveloped to large-scale institutional factors such as the new decision criteria in a process of environmental assessment. In this respect, the 'battle' between technocratic and hydrosocial water management will never be completely won or lost, though the weight of one approach's influence may wane and another may wax. It also follows that water management is just one field in which this battle is played out: parallel 'battles' (or, in fact, the same struggle) are underway within other technical fields such as energy management or transport planning.

Second, it is clear that language and understandings are important. In this respect, it would not be enough for hydrosocial water management to be hidden under a bushel and seen as (potentially) part of integrated water resource management or some other nirvana concept. If a more participative mode of working is to be applied in a more systematic way across the world's water systems, then it needs to be talked about and promoted explicitly. Linton has begun this process in an academic mode through introducing the concept of hydrosocial water management (Linton, 2014; Linton and Budds, 2014). In exploring in more depth what 'hydrosocial' means for the conceptualisation and practice of water management, I hope this book makes a contribution to the concept being better known and more widely and explicitly discussed.

Third, beyond language and understandings there is a need to institutionalise hydrosocial water management. This means a variety of changes including, but not confined to, those shown below.

- Training and gaining employment for a new cadre of hydrosocial water managers, equipped with broad technical understandings about how the water system has evolved and functions, knowledge of how socio-technical systems change and specific skills and knowledge about mobilisation and participation processes, in order to support water organisations in understanding and implementing a more hydrosocial approach.
- Training of a new cadre of engineers to have a basic understanding of policy-making processes and hence to be able to accept and value alternative perspectives on a technical system without dismissing them as invalid.
- Knowledge of the concept of hydrosocial water management by many water managers and stakeholders across the water sector and active debate about what the concept means and how it might be applied in specific incidences.
- Development of new sub-academic disciplines focused on hydrosocial water management, researching questions of whether and how water systems are evolving to embrace a hydrosocial approach. Specifically, in developing best practice in mobilisation and participation techniques there is a need for academic rigour in assessing and supporting water.
- Expectations about processes of participation and mobilisation being embedded in water management rules and procedures.

Routes to further embed the hydrosocial approaches

This section addresses the specific challenges that are faced by hydrosocial water management, and how they might be addressed.

Promote 'hydrosocial water management'

Changing the 'language' through which we talk about water is not a goal in itself: rather, it is a means through which changes in policy can be achieved. As noted above, there are various existing concepts that – in many interpretations – already work towards hydrosocial water management; however, because these concepts also have valid technocratic interpretations, to expect a further shift towards a more hydrosocial approach to be accomplished without changes in the language and conceptualisation of water management is naive. In this respect, the term 'hydrosocial water management' (or something similar conveying the same processes of reconnecting people with water) is an important component in helping a change to occur.

How can we ensure that water-related people read, write and talk about hydrosocial water management more? Mostly, the answer is by writing and talking about it ourselves!

So, as a reader of this book, *if* the concept makes sense to you, *if* you think the implicit values are ones that you want to promote, then it is over to you to go and use the concepts, to talk about it, to promote it and the changes it discusses in your local contexts, and to enable change to occur.

Valuing social information in water organisations

How can we shift understandings to give weight to societal perspectives and preferences away from the emphasis on technical knowledge that water utilities (and the public!) are accustomed to? It should be stressed that the hydrosocial approach in no way rejects engineering and ecological knowledge; the claim is rather that these forms of information need to be complemented by social knowledge and social understandings. Effectively, this perspective argues that there is more than one type of valid and useful knowledge, and hence that we need 'knowledges' to make good decisions.

Real progress is already being made in this field, particularly in the recognition that public and lay voices are relevant to landscape decisions. In some organisations social science knowledge is also sought and accepted. Further progress in this field suggests a need to focus on:

- the education of engineers in terms of the relevance and usefulness of social knowledge to some decisions and actions
- the education of a cadre of social scientists who have sufficient knowledge of water processes to be useful and authoritative in feeding into water-related decisions and processes
- improving the quantity and quality of social research relating to the practice of water management, and particularly to processes of engagement.

The impact of valuing social knowledge is that water organisation will be more able to interact with their publics in ways that are respectful, that value the public's knowledge and that develop and draw on this to build new ways of managing water together.

Reconnecting with water

As this book has highlighted, it is hard to involve the public in water-related decisions because there is a general lack of understanding about the urban parts of the water cycle. Examples of the 'facts' that some may have found surprising in this book are: the information that most urban water is consumed in 'virtual' form with food and through manufactured goods, not directly as water; that the toilet is the domestic device consuming most domestic water at present, though the rising expectations about the cleanliness of clothes means that the washing machine is a close second; and that it is possible for an urban area to have a satisfactory decentralised water supply and waste water system. If people are going to be in a position to give useful input to water-related decisions then some of these general water-related understandings may need to be more widely spread; understandings are also likely to be further enhanced if people have better knowledge of their household's water use and the local water situation to which local households, schools and businesses are contributing.

Already, many water organisations provide learning materials to schools to support science and geography lessons. This book suggests that this material might appropriately be extended to encompass more social aspects of water management – who uses what, where, how and, crucially, what difficult choices are faced moving forward. The material might also contribute to history curricula. More excitingly, modern information systems provide the means to offer much more local information which, supported by appropriate help with respect to interpretation, could provide school children with opportunities for science and geography projects that seek to develop good local understandings of how water flow in particular locations changes in quantity and quality through time.

Finally, and perhaps most importantly, 'reconnecting people with water' is most effectively achieved in relation to real choices faced in terms of how the system should develop. This relates to the process of water organisations moving beyond the legacy of nineteenth-century physical systems, and is therefore discussed in the next section.

Moving beyond legacy systems

Infrastructure systems are notoriously hard to provide because they require long-term investment. These can be hard to justify, particularly in the context of an otherwise restricted budget. Yet, whether they develop in a technocratic or social way, there is little doubt that the industrialised world's water systems are ageing and that significant new investment is required to make them more effective, less risk-prone and cheaper to operate. There are difficult decisions over which are the priority investments, who should make them and when they should happen. It is, therefore, highly appropriate that water utilities should reach out and seek to communicate to their publics their choices and dilemmas about the future. Of course, the full complexity of decision-making cannot be presented, but some of the choices and their positive and negative consequences can be put forward. In this respect, rather than assuming that the water organisation has first to do an option appraisal and select the best route forwards before consulting with the public, water organisations should, instead, get straight out there and explain choices and dilemmas they face, garnering people's opinions and, at the same time, increasing their knowledge of the system. Of course, many people will not engage, but, for those that do, thinking about how the system works and what sorts of changes will have what impact on them is a highly effective way of raising people's consciousness of water as a collective resource. By thinking in the long term, and by presenting the choices between different options for developing the system, the debate is focused on what is needed rather than just on the cost of investment.

Crossing system boundaries

One of the big criticisms of technocratic water management system has been that it worked in siloes. Some of these siloes concerned different parts of the water sectors – for example, flooding and water quality – but the water sector as a whole was also in a silo from other infrastructure sectors and, indeed, other social priorities, such as education and social care. Nevertheless, as highlighted in Chapter 3, it is not possible to integrate everything. That chapter charted the compromises in the UK, with the integration of functions within a locality giving way to the need to integrate spatially across the catchment. The story raises a question about whether and how it is possible to move beyond siloed working of different functions without giving up the advantages (e.g. availability of technical expertise) of catchment or regionally based geographical integration.

This question of 'how do we integrate across functions and geographies' is faced in many different policy sectors. It is certainly not easy. A first step might be achieved through research that could:

- assess the co-benefits and co-costs that occur at the boundaries between different systems – for example, between flood and water management or between water management and energy management
- identify where, when and how particular successful integrations (e.g. between flood management and water quality management) have occurred in the water sector and evaluate these
- collate learning from other policy sectors, identifying whether and how collaborations between sectors have worked and how they have been achieved.

In the longer run, there is a need to think about larger-scale institutional changes. There is certainly a case for developing funding that promotes a variety of water-related goals together rather than having separate funding streams that concern each goal individually. This approach, however, goes against the pattern of government budget-making that allocates particular funds to particular priorities. In the first instance, a small fund that seeks to support projects that 'integrate' functions could support the development of knowledge and expertise to support the latter development of an integrated fund. To some extent such integration between policy areas is already underway in research funding (for example, in supporting work on the 'nexus'), but it is important that such integration is rolled out to support processes of development and practice.

Other processes (some of which already occur) that could support decision-making across traditional policy field boundaries include:

- use of the municipality as the basic 'unit' of a locality and which is therefore seen as having expertise in a variety of local needs and their integration

(however, as noted above, this should not be exclusive, or it creates boundary issues of its own)

- prizes for investment schemes that bridge agendas, or, similarly, a small fund that is intended to support projects that achieve multifunctional goals rather than serving one agenda or another
- communication and joint project working between stakeholders involved in different sets of water infrastructure.

Finding ways to deal with trade-offs in relation to competing values

Given the recognition that scientific knowledge can be complemented by other types of knowledge, including local knowledge and [interpretive] social scientific knowledge, it becomes possible and, indeed, likely that, at times, different forms of knowledge will suggest different routes forward. The case in this book that most aptly illustrated stakeholder disagreement was the example of flooding the Somerset Levels. There were strong differences between the values and suggested actions advocated by environmental conservationists, on the one hand, and farmers, on the other. So how should hydrosocial water management deal with conflicts about water management actions in the present and the future?

The key lesson from this book is that such dilemmas should be discussed rather than hidden. Being 'out and proud' about a dilemma provides the means to stimulate public discussion. It enables the values underlying different choices to be explored. It also provides space for non-water-related concerns and knock-on effects of the decision to be aired. At its best, sharing a process of decision-making with the public enables a reconceptualisation and sharing of understandings and interests which allows mutual agreement on what should be done. To ensure that it achieves these high standards, water organisations need to employ high-quality managers who can plan and organise effective public communications and participation processes. Insights as to how to deal with differences and how to move forward when stakeholder groups fall into polarised camps can be obtained from experiences of stakeholder disputes in other fields.

Effectively, the call is for water managers to be clear about the political nature of the decisions they make. For many water engineers the thought of making water systems more political will cause them to shudder. Already, they may argue, they are subject to enough politics and the unwillingness of municipalities or regulators to provide investment and to recognise any problems in the systems that are not causing an immediate hazard. To make the decisions *more* explicitly political (they might think) is surely to subject themselves even more to the whims and changing priorities of party politics and public opinion.

The answer to this issue is to stress, first, that water is political with a small 'p' (in that water decisions do have inherent implications about values and priorities), but mostly not political with a big 'P'. Being open about the dilemmas and choices is a route to making decisions together rather than separately.

In the case of flooding investment in the Somerset Levels, interestingly, there *is* a semi-participatory process that determines the regional priorities for flooding investment in the context of broader governmental priorities. Contrary to media reports at the time, the decision-making was not entirely that of a technical agency imposing their views on a reluctant public. It is symptomatic of the default technical understanding of flood management that this process is generally hidden from view, and that overall government priorities for flood management are subject to little debate or discussion. The consequence – which was unfortunate for the technical agency and the government – was that there was no mention of this process or its rationale in media coverage of the floods.

Contribution to social science

This chapter has been largely focused on summarising the development of hydrosocial water management. As such, the discussion has illustrated and developed how social science speaks to the currently more technical worlds of water engineering and water ecology. As argued in the Introduction, however, the book also aims to speak to a broader social science community. Its purpose is not to offer a detailed development of social science theory about water, but rather to synthesise, elaborate and bring together a wide body of existing work. In terms of the book's contribution to social science, five contributions can be identified.

First, it provides a broad understanding of water management. Many social scientists may be unfamiliar with the physical and human systems through which waters are channelled, or with elements of their history and operation. This book offers a summary of how water management has developed as it has, demonstrating something of its current operation and describing some of the recent social science research on these topics. Specifically, it has offered a political interpretation that shows technocratic water management beginning to give way to a more hydrosocial approach.

Second, the book has elaborated the concept of hydrosocial water management in the context of contemporary water management practice. The term 'hydrosocial water management' originates with Linton and a set of human geographers who have used political ecology to produce some useful critical commentaries about water management. Through the chapters of this book I have sought to link this concept into some detailed examples of processes of change in water management, and also to understand the changes in the context of some social science theory, including practice theory and Douglas' cultural theory. If hydrosocial water management is becoming more prevalent, as suggested, it indicates that there will be much more call for social science in water in the future.

Third, I have offered an interpretation of how hydrosocial water management can be seen in broader governance terms. Through the use of Mary Douglas' cultural theory, the book has offered a commentary on change in the field of water management that might be applicable in other socio-technical fields such as energy management or transport planning. Specifically, there is a suggestion that the traditional 'technocratic' approach provides hierarchical governance that is under pressure to change. While some commentaries and approaches stress the need for more monetisation of decision-making associated with individualistic governance, this book has highlighted instead the possibilities for more egalitarian governance through hydrosocial water management.

Fourth, this book offers some pointers as to where interpretive social science research on water should go next. At the heart of the research agenda arising from this volume is the need for social and political perspectives on water management that are accessible to practitioners and policy-makers beyond the academy. An 'accessible social science of water' is required if these perspectives are to have a chance of influencing water management. It should be stressed that this call does not mean that social scientists of water abandon all communication with their disciplinary community – far from it. Rather, the argument is that they (or at least, some of us) need to work to build bridges from academic social science to the worlds of water policy and practice. As this volume has demonstrated, by doing so we can not only offer new understandings about water governance and practice, but we can also contribute to broader social science theory and debates – for example, in relation to changing governance processes as discussed in Chapter 8.

Within this general category of accessible interpretive social science of water, some specific research needs can be discerned. As Chapters 5 and 9 have shown, there is a need for better-developed and more broadly accepted processes for analysing and evaluating mobilisation and participation activities in water services. These are likely to involve a combination of short- and medium-term 'process' goals and longer-term outcome-related goals (Carr *et al.*, 2012). Another area of need concerns processes of technical innovation; social scientists need to be involved in proposals for technical innovation (of water, but also of other things) from an early stage in terms of helping to anticipate their impacts, and so that processes of participative design and development can be developed and supported. A third area where there is relatively little investment of contemporary research effort concerns processes of social or governance innovation in relation to water. Whether bridging between sectors (e.g. between flood management and water quality management) inside or outside their organisations, or whether developing new relationships with their stakeholders, such changes in the social arrangements for water services need to be studied and evaluated just as much as the equivalent technical innovations. Without such evaluation, how can learning occur? Finally, but perhaps combining with all of the above, there is also a need for more collaborative research with water organisations – whether utilities, regulators or civic society organisations – to better understand, challenge, support and develop actions to improve their water services (Westling *et al.*, 2014). Such research should help to yield more empirical evidence about the variety of social science contributions to water services, but will also deliver insights on the nature of social science expertise and its potential to aid the transition of a previously technocratic service to something which is (one hopes) more open to the needs of its publics.

This issue, the potential for interpretive social science to help water management develop, links to the book's final contribution to social science. I hope that it has provided an ambitious example of social science engaging with a technical subject. As argued in Chapter 1, social science engagement with technical subjects tends to be either at the level of detached political commentary or the elucidation of detailed case studies. It is almost as if some social scientists are scared of the technical detail of a subject like water management, and of the thought that their understandings will not be valued. Through engaging broadly with water management science and social science to bring the critical commentaries and detailed case studies together, this book has offered an example of how social scientists might constructively engage with technical topics. The implicit point is that the detached comments are *too* detached from the cut and thrust of decision-making, while detailed narratives and case studies are valuable – but can offer even more value when brought together and seen as part of a broader process of change. By articulating what this process of change is and how it manifests across different spheres of water management, I hope to provide insight to practitioners and academics from both engineering and social disciplines alike. The particular contribution that would be most satisfying is if, through reading this work, people are inspired to reflect on and to develop their own practices. In completing this book, my aspiration is that it will help water managers and engineering researchers to see their own work in a new light while also inspiring more social scientists to venture into the complex and confusing world of a technical subject secure in the knowledge that they too can produce research which is of use.

REFERENCES

Acreman MC, Harding RJ, Lloyd C, McNamara NP, Mountford JO, Mould DJ, Purse BV, Heard MS, Stratford CJ and Dury SJ, 2011, Trade-offs in ecosystem services of the Somerset Levels and Moors wetlands. *Hydrological Sciences Journal*, 56 (8): 1543–65

Aire Rivers Trust, 2012, *Bradford's Becks: A New Lease of Life*, available online at https://bradford-beck.org/catchment-management-plan/, last accessed 15 August 2016

Allan JA, 2003, Virtual water – the water food and trade nexus: useful concept or misleading metaphor, *Water International*, 28: 1

Allen M, 2008, *Cleansing the City: Sanitary Geographies in Victorian London*, Athens, OH: Ohio University Press

Allon F and Sofoulis Z, 2006, Everyday water: cultures in transition, *Australian Geographer*, 37: 151–79

Anon, 2014, *Somerset Levels and Moors Flood Action Plan* (March 2014), available online at https://somersetnewsroom.com/flood-action-plan/, last accessed 15 August 2016

Aquality, 2016, *WWF HQ Sustainability*, available online at aqua-lity.co.uk, last accessed 29 June 2016

Arnstein S, 1969, A ladder of citizen participation, *Journal of the American Institute of Planners*, 34: 216–25

Ashenburg K, 2009, *Clean: An Unsanitised History of Washing*, London: Profile

Bakker K, 2004, *An Uncooperative Commodity: Privatizing Water in England and Water*, Oxford: Oxford University Press

Bakker K, 2014, The business of water: market environmentalism in the water sector, *The Annual Review of Environment and Resources*, 39: 469–94, doi: 10.1146/annurev-environ-070312-132730

Bell S, 2015, The place that will save us when there's a drought: London's desalination plant, *The Londonist*, October, last accessed 13 May 2016

Bell S and Aitken V, 2008, The socio-technology of indirect potable water reuse, *Water Science and Technology*, 8 (4): 441–8, doi: 10.2166/ws.2008.104

Berger J, 1972, *Ways of Seeing*, London: Penguin

Binney P, Donald A, Elmer V, Ewert J, Phillis O, Skinner R and Young R, 2010, *IWA Cities of the Future Program: Spatial Planning and Institutional Reform Discussion Paper for the World*

Water Congress, September, International Water Association, available online at www.clearwater.asn.au/user-data/case-studies/plans-designs/Conclusions-Cities-of-the-Future-Montreal_2011.pdf, last accessed 11 July 2016

Biswas AK, 2004, Integrated water resource management: a reassessment: a water forum contribution, *Water International*, 29 (2): 248–56

Brown A and Matlock MD, 2011, *A Review of Water Scarcity Indicators and Methodologies*, The Sustainability Consortium: University of Arkansas, available online at www.sustainabilityconsortium.org, last accessed 17 September 2013

Brown RR and Clarke J, 2007, *Transition to Water Sensitive Urban Design: The Story of Melbourne, Australia*, Facility for Advanced Biofiltration: Monash University, June, ISBN 978-0-9803428-0-2

Brown RR and Farrelly MA, 2009, Delivering sustainable urban water management: a review of the hurdles we face, *Water Science and Technology*, 59 (5): 829–46

Brown RR, Sharp L and Ashley RM, 2006, Implementation impediments to institutionalising the practice of sustainable urban water management, *Water Science and Technology*, 54 (6–7): 415–22

Brown RR, Keath N and Wong THF, 2009, Urban water management in cities: historical, current and future regimes. *Water Science and Technology*, 59 (5): 847–55

Browne AL, 2015, *Dirt, Cleanliness, Everyday Life and Festivals: 'Traditional' Methodologies as Sites of Experimentation for Sustainable Water Consumption Agendas*, presentation at 'HomeLabs: testing promising practices for sustainable household consumption', CONSENSUS project 2nd International Conference, Galway, University of Ireland, 21–22 May 2015

Browne AL, Medd W and Anderson B, 2013a, Developing novel approach to tracing domestic water demand under uncertainty: a reflection on the 'up scaling' of social science approaches in the United Kingdom, *Water Resource Management*, 27 (4): 1013–35, doi: 10.1007/s11269-012-0117-y

Browne AL, Pullinger M, Medd W and Anderson B, 2013b, *Patterns of Water: Resource Pack*. Lancaster: Lancaster University

Bryson B, 2010, *At Home: A Short History of Private Life*, London: Transworld

Burian SJ, Nix S, Pitt R and Durrans SR, 2000, Urban wastewater management in the United States: past, present and future, *Journal of Urban Technology*, 7 (3): 33–62

Business Environment Network, 2011, *Stormwater Harvesting for Melbourne Park*, available online at http://ben.uat.aspermont.com/storyview.asp?StoryID=9586688&redirect=commentclick&commentlevel=2§ionsource=&aspdsc=yes, last accessed 12 July 2016

Butler D and Memon FA, 2006, *Water Demand Management*, London: IWA

Butler C, and Pidgeon N, 2011, From 'flood defence' to 'flood risk management': exploring governance, responsibility, and blame, *Environment and Planning C: Government and Policy*, 29(3): 533–47

Cairns R and Krzywoszynska A, 2016, Anatomy of a buzzword: the emergence of 'the water-energy-food bexus' in UK natural resource debates, *Environmental Science and Policy*, 64: 164–70

Carr G, Blöschl G and Loucks DP, 2012, Evaluating participation in water resource management: a review, *Water Resources Research*, 48, W11401: 1–17, doi:10.1029/2011WR011662

CCW [Consumer Council for Water], 2015a, *Average Water Use*, available online at www.ccwater.org.uk, last accessed 10 February 2015

CCW, 2015b, *Delving into Water 2015: Performance of Water Companies in England and Wales 2010–11 and 2014–15*, December, available online at www.ccwater.org.uk/wp-content/uploads/2015/12/Delving-into-water-2015-FINAL.pdf

Cettner A, 2012, Overcoming inertia to sustainable stormwater management practice, doctoral thesis, University of Luleo, Sweden

Chappels H and Medd W, 2008, From big solutions to small practices: bringing back the active consumer, *Social Alternatives*, 27 (3): 44–9

Chew MYC, Watanabe C and Tou Y, 2011, The challenges in Singapore NEWater development: co-evolutionary development for innovation and industry evolution, *Technology in Society*, 33 (3–4): 200–11, doi:10.1016/j.techsoc.2011.06.001

Chilvers J and Evans J, 2009, Understanding networks at the science–policy interface, *Geoforum*, 40 (3): 355–62, doi: 10.1016/j.geoforum.2009.03.007

City of York, 2012, *Surface Water Management Plan*, available online at www.york.gov.uk, last accessed 15 August 2014

City of York, 2013, *Strategic Flood Risk Assessment*, available online at www.york.gov.uk, last accessed 15 August 2014

Collins H and Evans R, 2002, Third wave of science studies: studies of expertise and experience, *Social Studies of Science*, 32: 235–96, doi: 10.1177/0306312702032002003

Committee on Climate Change, 2014, *Policy Note: Flood and Coastal Erosion Risk Management Spending*, available online at www.theccc.org.uk/publication/policy-note-flood-and-coastal-erosion-risk-management-spending/, last accessed 28 July 2016

Comprehensive Assessment of Water Management in Agriculture, 2007, *Water for Food, Water for Life: A Comprehensive Assessment of Water Management in Agriculture*, London: Earthscan

Connelly S and Anderson C, 2007, Studying water: reflections on the problems and possibilities of inter-disciplinary working, *Interdisciplinary Science Reviews*, 32: 213–32

Connelly S and Richardson T, 2004, Exclusion: the necessary difference between ideal and practical consensus, *Journal of Environmental Planning and Management*, 47 (1): 3–17

Cook BR and Spray CJ, 2012, Ecosystem services and integrated water resource management: different paths to the same end? *Journal of Environmental Management*, 109, 93–100

Cornwall A and Gaventa J, 2000, From users and chooser to makers and shapers, *IDS Bulletin*, 31 (4): 50–62

Cornwall A and Coelho VSP, 2007, Spaces for change? The politics of participation in new democratic arenas, in A Cornwall and VSP Coelho, *Spaces for Change? The Politics of Citizen Participation in New Democratic Arenas*, London and New York: Zed, pp. 1–29

Costanza R, d'Arge R, de Groot R, Farber S, Grasso M, Hannon B, Limburg K, Naeem S, O'Neill RV, Paruelo J, Raskin RG, Sutton P and van den Belt M, 1997, The value of the world's ecosystem services and natural capital, *Nature*, 387: 253–60

Cotton M, 2016, Fair fracking? Ethics and environmental justice in UK shale gas policy and planning, *Local Environment*, doi: 10.1080/13549839.2016.1186613

Darcy S, 2016, *Stakeholder Perspectives on Water Market Reform (Sustainability First)*, speech given to 'New steps for water market reform in England: competition, infrastructure and priorities for the future' conference, Westminster Energy, Environment and Transport Forum Keynote Seminar, 5 July 2016

DCWW, 2016, *Rainscape*, available online at www.dwrcymru.com/en/My-Wastewater/RainScape.aspx, last accessed 9 March 2016

de Graaf R, 2009, Innovations in urban water management to reduce the vulnerability of cities, thesis from Delft University of Technology, Delft

Defra (Department for Environment, Farming and Rural Affairs), 2008, *Future Water: The Government's Water Strategy for England*, Cmnd 7319, available online at www.gov.uk/government/publications/future-water-the-government-s-water-strategy-for-england, last accessed 8 October 2013

Defra 2009 *Public Understanding of Sustainable Water use in the Home: A Research Report Completed for the Dept for Environment, Food and Rural Affairs* by Synovate, London: Defra, available online at www.gov.uk, last accessed 4 April 2016

Defra, 2010, *Water Retrofit Policies Review: The Household Perspective*, R&D Technical Report FD2649/TR, London: Defra

Defra, 2011, *Upland Policy Review*, available online at www.gov.uk, last accessed 14 August 2014

Doron U, Teh T, Haklay M and Bell S, 2011, Public engagement with water conservation in London, *Water and Environment Journal,* 25: 555–62

Douglas M, 1984, *Purity and Danger: An Analysis of the Concepts of Pollution and Taboo*, London: Ark

Douglas M, 1985, *Risk Acceptability According to the Social Sciences*, New York: Russell Sage Foundation

Douglas M and Wildavski A, 1982, *Risk and Culture: An Essay on the Selection of Technological and Environmental Dangers*, Berkeley: University of California Press

Douglas M, Thompson M and Verweij M, 2003, Is time running out? The case of global warming, *Daedalus*, 132 (2): 98–107

Downs A, 1972, Up and down with ecology: the 'issue-attention cycle', *National Affairs*, 28: 38–50

EA (Environment Agency), 2007, *Review of the 2007 Summer Floods*, Bristol: Environment Agency, available online at www.environment-agency.gov.uk, last accessed 10 November 2014

EA, 2008a, *Water Resources in England and Wales: Current State and Future Pressures*, Bristol: Environment Agency, available online at www.environment-agency.gov.uk, last accessed 19 September 2013

EA, 2008b, *International Comparison of Domestic Per Capita Consumption*, prepared by Aquaterra, Reference: L219/B5/6000/025b, available online at www.environment-agency.gov.uk, last accessed 17 September 2013

EA, 2010, *River Ouse Catchment Flood Management Plan*, July 2010, available online at www.gov.uk, last accessed 15 August 2014

EA, 2016, *Risk of Flooding from Rivers and Sea*, available online at http://watermaps.environment-agency.gov.uk/, last accessed 24 July 2016

EPA (Environmental Protection Authority), 2016, *Soak Up the Rain*, available online at www.epa.gov/soakuptherain, last accessed 24 July 2016

European Parliament and Council, 2007, *DIRECTIVE 2007/60/EC of 23 October 2007 on the assessment and management of flood risks*, available online at http://ec.europa.eu/environment/water/flood_risk/, last accessed 15 August 2016

Eyre A, 1875, A Dyurnall, or catalogue of all my accions and expences, *Surtees Society*, 65: 1–118

Falkenmark M, 1989, The massive water scarcity threatening Africa: why isn't it being addressed? *Ambio*, 18 (2): 112–18

Falkenmark M, 2008, Water and sustainability: a reappraisal, *Environment*, 50 (2): 4–16

Fam D and Lopes AM, 2015, Designing for system change: innovation, practice and everyday water, *ACME*, 14 (3): 751–64

Fam D, Lahiri-Dutt K and Sofoulis Z, 2015, Scaling down: researching household water practices, *ACME*, 14 (3): 639–51

FAO, 2014, Aquastat Home, available online at www.fao.org/nr/water/aquastat/main/index.stm, last accessed 7 December 2014

Fernandez-Gimenez ME, Ballard HL and Sturtevant VE, 2008, Adaptive management and social learning in collaborative and community-based monitoring: a study of five community-based forestry organizations in the western USA, *Ecology and Society*, 13 (2): 4

Fischer F, 2009, *Democracy and Expertise*, Oxford: Oxford University Press

Fisher J, Cotton AP and Reed BJ, 2005, *Learning from the Past: Delivery of Water and Sanitation Services to the Poor in 19th Century Britain*, Background Report for WELL Briefing Note 9, Leicestershire: Loughborough University

Fisher B, Turner RK and Morling P, 2009, Defining and classifying ecosystem services for decision making, *Ecological Economics*, 68: 643–53

Flyvbjerg B, 2001, *Making Social Science Matter: Why Social Inquiry Fails and How it can Succeed Again*, Cambridge: Cambridge University Press

Folke C, 2006, Resilience: the emergence of a perspective for social–ecological systems analyses, *Global Environmental Change*, 16: 253–67

Foster J. (ed.), 2003, *Valuing Nature: Economics, Ethics and the Environment* (2nd edn), London: Routledge

Foucault M, 1977, *Discipline and Punish*, translated by Alan Sheridan, London: Vintage

Gabe J, Trowsdale S and Mistry D, 2012, Mandatory urban rainwater harvesting: learning from experience, *Water Science and Technology*, 65 (7): 1200–7

GARD (Group Against Reservoir Development), 2016, *Reservoir Plan Remains A Threat!* available online at www.abingdonreservoir.org.uk, last accessed 15 August 2016

Garis Y, Lutt N and Tag A, 2003, Stakeholder involvement in water resources planning, *Water and Environment Journal*, 17 (1): 54–8

Geels F, 2005, Co-evolution of technology and society: the transition in water supply and personal hygiene in the Netherlands (1850–1930): a case study in multi-level perspective, *Technology in Society*, 27: 363–97

Gleick PH, 1996 Basic water requirements for human activities: meeting basic needs, *Water International* (IWRA), 21: 83–92

Grunig JE and Hunt T, 1984, *Managing Public Relations*. New York: Holt, Rinehart and Winston

Green Facts, 2014, *Water Resources*, available online at www.greenfacts.org/en/water-resources/, last accessed 7 December 2014

GWP (Global Water Partnership), 2013, *What is IWRM?* available online at www.gwp.org/, last accessed 29 July 2013

GWP–TAC (Global Water Partnership – Technical Action Group), 2000, *Integrated Water Resource Management*, TAC Background Paper No. 4. GWP, Stockholm, Sweden

Hajer M, 1995, *The Politics of Environmental Discourse: Ecological Modernization and the Policy Process*, Oxford: Clarendon Press

Hajer M and Wagenaar H, 2003, Introduction, in Hajer J and Wagenaar H (eds), *Deliberative Policy Analysis*, Cambridge: Cambridge University Press, pp. 1–32

Halliday S, 1999, *The Great Stink of London: Sir Joseph Bazalgette and the Cleansing of the Victorian Metropolis*, Stroud: History Press

Hamlin C, 1988, Muddling in Bumbledom: on the enormity of large sanitary improvements in four British towns, 1855–1885, *Victorian Studies*, 32: 72–4

Hamlin C, 1992, Edwin Chadwick and the engineers, 1842–1854: systems and antisystems in the pipe-and-brick sewers war, *Technology and Culture*, 33 (4): 680–790

Hansen R, undated, Water related infrastructure in medieval London, available from http://www.waterhistory.org/histories/london/london.pdf, last accessed 9 April 2016

Haraway D, 1988, The science question in feminism and the privilege of partial perspective, *Feminist Studies*, 14 (30): 575–99

Hartley D, 1964, *Water in England,* London: Macdonald

Harvey D, 1996, *Justice, Nature and the Geography of Difference*, Chichester: Wiley-Blackwell

Haughton G, 2015, Knowledge, power and emotions in stakeholder participation within environmental governance, unpublished thesis, University of Sheffield Faculty of Social Sciences, Urban Studies and Planning

Head L, Gibson C, Gill N, Carr C and Waitt G, 2016, A meta-ethnography to synthesise household cultural research for climate change response, *Local Environment*, doi: 10.1080/13549839.2016.1139560

Healey P, 1997, *Collaborative Planning: Shaping Places in Fragmented Societies*, Basingstoke: Macmillan

Heems GC and Kothuis BLM, 2012, Flood safety: managing vulnerability beyond the myth of dry feet. A socio-cultural perspective on Dutch ways of coping with flood risk, PhD thesis (in Dutch), Amsterdam: Waterworks

Hey D, 2010, *A History of Sheffield* (3rd edn), Lancaster: Carnegie

Hicklin AJ, 2004, Flood management communities and the environment, *Water and Environment Journal*, 18 (1): 20–4

Hill M, 2013, *The Policy Process* (6th edn), Oxford: Routledge

Hoekstra AY (ed.), 2003, *Virtual Water Trade: Value of Water Research Report No 12*, Delft: UNESCO IHE

Hoekstra AY, 2014, Sustainable, efficient and equitable water use: the three pillars under wise fresh water allocation, *WIREs Water*, 1: 31–40. doi: 10.1002/wat2.1000

Hoekstra AY and Chapagain AK, 2007, Water footprints of nations: water use by people as a function of their consumption pattern, *Water Resources Management*, 21, 35–48

Hoff H, 2011, *Understanding the Nexus. Background Paper for the Bonn 2011 Conference: The Water, Energy and Food Security Nexus*, Stockholm: Stockholm Environment Institute

Hood C, 1998, *The Art of the State: Culture, Rhetoric, and Public Management*, Oxford: Clarendon

Hoskins WG, 2005, *The Making of the English Landscape*, London: Hodder & Stoughton [new edn]

House of Lords, 2006, *Water Management: 8th Report of Science and Technology Committee in the Sessions 2005–6*, available online at www.publications.parliament.uk, last accessed 28 March 2016

Hughes M, 2013, The Victorian London sanitation projects and the sanitation of projects, *International Journal of Project Management*, 31: 682–91

Imhof A, Wong S and Bosshard P, 2002, *Citizens' Guide to the World Commission on Dams*, available online at www.internationalrivers.org, last accessed 10 May 2016

International Rivers, 2016, *Our Work*, available online at www.internationalrivers.org/campaigns/the-world-commission-on-dams, last accessed 3 March 2016

Jacobs M, 1993, *The Green Economy, Environment, Sustainable Development and the Politics of the Future*, Toronto: University of Toronto Press

Jasanoff S, 2003, Technologies of humility: citizen participation in governing science, *Minerva*, 41: 223–44

Jeffrey P and Gearey M, 2006, Integrated water resources management: lost on the road from ambition to realisation? *Water, Science and Technology*, 35 (1): 1–8

Jeffries N, 2006, The Metropolis Local Management Act and the archaeology of sanitary reform in the London Borough of Lambeth 1856–86, *Post Medieval Archaeology*, 40 (1): 1–20

Jenner M, 2000, From conduit to commercial network? Water in London, 1500–1750, in P Griffiths and Jenner M (eds), *Londonopolis: Essays in the Cultural and Social History of Early Modern London*, Manchester: Manchester University Press, pp. 250–72

Johnson C, Penning-Rowsell E and Parker D, 2007, Natural and imposed injustices: the challenges in implementing 'fair' flood risk management policy in England, *The Geographical Journal*, 173 (4): 374–90

Johnson S, 2008, *The Ghost Map*, London: Penguin

Kaika M, 2006, The political ecology of water scarcity: the 1989–1991 Athenian drought, in N Hyeman, M Kaika and E Swyngedouw (eds), *In the Nature of Cities: Urban Ecology and the Politics of the Urban Metabolism*, Oxford: Routledge, pp. 157–73

Karvonen A, 2011, *Politics of Urban Runoff: Nature, Technology and the Sustainable City*, Cambridge, MA: MIT Press

Kayaga S, Calvert J and Sansom K, 2003, Paying for water services: effects of household characteristics, *Utilities Policy*, 11 (3): 123–32

Knamiller C, 2011, The impact of domestic water use cultures on water efficiency interventions in the south east of England: lessons for water demand management, unpublished thesis, Department of Geography and Environmental Science, University of Bradford

Knamiller CA and Sharp L, 2009, Issues of trust, fairness and efficacy: a qualitative study of information provision for newly metered households in England, *Water Science and Technology: Water Supply*, 9: 323–31

Kothuis B, 2016, *Beyond the Myth of Dry Feet*, presentation at Delft University of Technology to students of Urban Studies and Planning (University of Sheffield), 16 March

Lane SN, 2003, More floods, less rain? Chapter in changing hydrology in a Yorkshire context, in M Atherden (ed.), *Global Warming in a Yorkshire Context*, available online at www.therrc.co.uk/publications/more-floods-less-rain-changing-hydrology-yorkshire-context-chapter-global-warming, last accessed 24 July 2016

Lane SN, Odoni N, Landstrom C, Whatmore SJ, Ward N and Bradley S, 2011, Doing flood risk science differently: an experiment in radical scientific method, *Transactions of the Institute of British Geographers*, NS 36: 15–36

Lee SJ, 1994, *Aspects of British Political History, 1815–1914*, London: Routledge

Lerner D, 2015, personal communication, 24 May

Linton J, 2004, Global hydrology and the construction of a water crises, *The Great Lakes Geographer*, 11 (2): 1–13

Linton J, 2008, Is the hydrologic cycle sustainable? A historical-geographical critique of a modern concept, *Annals of the Association of American Geographers*, 98 (3): 630–49

Linton J, 2014, Modern water and its discontents: a history of hydro-social renewal, *WIREs Water*, 1: 111–20. doi: 10.1002/wat2.1009

Linton J and Budds J, 2014, Defining and mobilizing a relational-dialectical approach to water, *Geoforum*, 57: 170–80

Lowe P and Goyder J, 1983, *Environment Groups in Politics*, London: Allen & Unwin

Mankad A and Gardner J, 2016, The role of personal experience in frequency of rainwater tank maintenance and policy implications, *Local Environment*, 21 (3): 330–43, doi: 10.1080/13549839.2014.959907

Medema W, McIntosh BS and Jeffrey PJ, 2008, From premise to practice: a critical assessment of integrated water resources management and adaptive approaches in the water sector, *Ecology and Society*, 13 (2): 29: 1–18

Mekonnen MM and Hoekstra AY, 2011, *National Water Footprint Accounts: The Green, Blue and Grey Water Footprint of Production and Consumption, Volume 1: Main Report*, UNESCO-IHE, Research Report Series No. 50, available online at www.waterfootprint.org/Reports/Report50-NationalWaterFootprints-Vol1.pdf, last accessed 30 September 2013

Mitchell VG, 2006, Applying integrated urban water management concepts: a review of Australian experience, *Environmental Management*, 37 (5): 589–605

Molle F, 2008, Nirvana concepts, narratives and policy models: insight from the water sector, *Water Alternatives*, 1 (1): 131–56

Molyneux-Hodgson S, 2015, personal communication, 25 June

Moore S, 2011, Parchedness, politics, and power: the state hydraulic in Yemen, *Journal of Political Ecology*, 18: 39–50

Mortimer I, 2008, *The Time Traveller's Guide to Medieval England*, London: Random House

Mortimer I, 2013, *The Time Traveller's Guide to Elizabethan England*, London: Random House

Murphy S, 2014, *Utilities of the Future: Performance and Sustainability*, keynote speech at IWA Lisbon Congress, 21–26 September

Nevarez L, 1996, Just wait until there's a drought: mediating environmental crises for urban growth, *Antipode*, 28: 246–72

Norgaard RB, 2010, Ecosystem services: from eye-opening metaphor to complexity blinder, *Ecological Economics*, 69, 1219–27

Norton M, 2014, Water security: pipe dream or reality? A global perspective from the UK, *WIREs Water* 1: 11–18. doi: 10.1002/wat2.1005

Nye M, TapselL S and Twigger-Ross C, 2011, New social directions in UK flood risk management: moving towards flood risk citizenship, *Journal of Flood Risk Management*, 4: 288–97

OFWAT (undated) Appendix 1: *Water efficiency targets 2010–11 to 2014–15*, available online at www.ofwat.gov.uk, last accessed 12 February 2015

OFWAT, 2011, *Exploring the Costs and Benefits of Faster, More Systematic Water Metering in England and Wales*, available online at www.ofwat.gov.uk, last accessed 3 July 2013

OFWAT, 2016, *Leakage*, available online at www.ofwat.gov.uk/households/supply-and-standards/leakage/, last accessed 15 August 2016

Oki T and Kanae S, 2006, Hydrological cycles and global water resources, *Science*, 313: 1068–72, doi: 10.1126/science.1128845

Olsson P, Folke C and Berkes F, 2004, Adaptive co-management for building resilience in social-ecological systems, *Environmental Management*, 34 (1): 75–90

Opinion Leader, 2006, *Using Water Wisely*, London: Consumer Council for Water (CCW)

Owens S, 2005, Commentary: making a difference? Some perspectives on environmental research and policy, *Transactions of the Institute of British Geographers*, 30 (3): 287–92

Oreskes N, 2004, Science and public policy: what has proof got to do with it? *Environmental Science and Policy*, 7: 369–83

Oxford Mail, 2016, *Controversial Giant Reservoir Planned for South of Abingdon is 'Inevitable', Claims Water Experts*, 21 January, available online at www.oxfordmail.co.uk, last accessed 10 May 2016

Page B, 2005, Paying for water and the geography of commodities, *Transactions of the Institute of British Geographers*, NS 30: 293–306

Page B and Bakker K, 2005, Water governance and water users in a privatised water industry: participation in policy-making and in water services provision: a case study of England and Wales, *International Journal of Water*, 3 (1): 38–60

Pahl-Wostl C, 2007, Transitions towards adaptive management of water facing climate and global change, *Water Resource Management*, 21: 49–62

Pahl-Wostl C, Newig J and Ridder D, 2008, Linking public participation to adaptive management, in P. Quevauviller (ed.), *Groundwater Science and Policy: An International Overview*, London: RSC

Patel M, 2008, The long haul: integrating water, sewage, public health and city building, in W Reeves and C Palassio, *H_2O: Toronto's Water*, Toronto: Coach House

Pearce D and Barbier E, 2000, *Blueprint for a Sustainable Economy*, London: Earthscan

Pitt M, 2008, *Learning the Lessons from the 2007 floods*, available online at http://webarchive.nationalarchives.gov.uk/, last accessed 15 August 2016

Pullinger M, Browne AL, Anderson B and Medd W, 2013, *The Water Related Practices of Households in Southern England and their Influence on Water Consumption and Demand Management*, final report of the ARCC Water/SPRG Patterns of Water projects, Lancaster University, available online at www.escholar.manchester.ac.uk/uk-ac-man-scw:187780, last accessed 1 February 2015

Reed MS, 2008, Stakeholder participation for environmental management: a literature review, *Biological Conservation*, 141 (10): 2417–31

Reynolds J, 2016, *Stakeholder Perspectives on Water Market Reform (Castle Water)*, speech given to 'New steps for water market reform in England: competition, infrastructure and priorities for the future' conference, Westminster Energy, Environment and Transport Forum Keynote Seminar, 5 July

Richter BD, Abell D, Bacha E, Brauman K, Calos S, Cohn A, Disla C, Friedlander O'Brien S, Hodges D, Kaiser S, Louhran M, Mestre C, Reardon M and Siefried E, 2013, Tapped out: how can cities secure their water future? *Water Policy*, 15: 335–63

Rijsberman FR, 2006, Water scarcity: fact or fiction? *Agricultural Water Management*, 80 (1–3): 5–22

Rivers Trust, 2014, *The Aire Rivers Trust*, available online at www.theriverstrust.org/riverstrusts/aire.html, last accessed 13 November 2014

Rouillard JJ, Heal KV, Ball T and Reeves AD, 2013, Policy integration for adaptive water governance: learning from Scotland's experience, *Environmental Science and Policy*, 33: 378–87

Rowe G and Frewer L, 2005, A typology of public engagement mechanisms, *Science, Technology, and Human Values*, 30 (2): 251–90

Rychlewski M, Westling EL, Sharp L, Tait S and Ashley R, 2013, *Adaptation Planning Process: Guidance Manual*, Bradford: University of Bradford, available online at www.sheffield.ac.uk/polopoly_fs/1.357472!/file/Adaptation_Planning_Process_Guidance_Manual.pdf, last accessed 15 January 2014

Savenije HHG and Van der Zaag P, 2008, Integrated water resources management: concepts and issues, *Physics and Chemistry of the Earth*, 33: 290–7

Sayer A, 1992, *Method in Social Science: A Realist Approach* (2nd edn), London: Routledge

Scarse JI and Sheate WR, 2005, Re-framing flood control in England and Wales, *Environmental Values*, 14: 113–37

Schmidt G and Benitz-Sanz C, 2013, *How to Distinguish Water Scarcity and Drought in EU Water Policy?* Global Water Forum, discussion paper 1333, available online at www.globalwater-forum.org/2013/08/26/how-to-distinguish-water-scarcity-and-drought-in-eu-water-policy/, last accessed 13 September 2013

Sefton C, 2008, Public engagement with sustainable water management, unpublished PhD thesis, University of Bradford

Sharp L, 2002, Participation and policy: unpacking connections in one UK LA21, *Local Environment*, 7 (1): 7–22

Sharp L, 2006, Water demand management in the UK: constructions of the domestic water user, *Journal of Environmental Management and Planning*, 49 (6): 869–89

Sharp L and Connelly S, 2002, Theorising participation: pulling down the ladder, in Y Rydin and A Thornley, *Planning in the UK: Agendas for the New Millennium*, London: Ashgate, pp. 33–63

Sharp L, Macrorie R and Turner A, 2015, Resource efficiency and the imagined public: insights from cultural theory, *Global Environmental Change*, 34, 196–206

Sharp, L, Bryant, M, Connelly, S and Moug, P, 2011a, Presentation at URSULA conference 2011, 'Participatory governance in action: the case of the Five Weirs Walk project', available at www.ursula.group.shef.ac.uk/upload/Inner/Outputs/URSULAConf2011/2011_Sharp_5WW.pdf, last accessed 11 January 2017

Sharp L, McDonald A, Sim P, Knamiller CA, Sefton C and Wong S, 2011b, Positivism, post-positivism and domestic water demand, *Transactions of the Institute of British Geographers*, NS 36: 501–15

Shove E, 2003, *Comfort Cleanliness and Convenience*, London: Berg

Shove E, 2010, Beyond the ABC: climate change policy and theories of social change, *Environment and Planning A*, 42: 1273–85

Sim P, McDonald A, Parsons J and Rees P, 2007 *Revised Options for UK Domestic Water Reduction: A Review*, University of Leeds WaND Briefing Note 28

Simmons G, Hope V, Lewis G, Whitmore J and Gao W, 2001, Contamination of potable roof-collected rainwater in Auckland, New Zealand, *Water Research*, 35 (6): 1581–24

Skelton L, 2016, *Sanitation in Urban Britain, 1560–1700*, London: Routledge

Smith A, Porter J and Upham P, 2016, 'We cannot let this happen again': reversing UK flood policy in response to the Somerset Levels floods 2015, *Journal of Environmental Planning and Management*, 60 (2): 351–69, doi: 10.1080/09640568.2016.1157458

Sofoulis Z, 2011, Skirting complexity: the retarding quest for the average water user, *Continuum: Journal of Media and Cultural Studies*, 25: 795–810

Sofoulis Z, 2015, The trouble with tanks: unsettling dominant Australian urban water management paradigms, *Local Environment*, 20 (5): 529–47, doi: 10.1080/13549839.2014.903912

Sofoulis Z, Allon F, Campbell M and Attwater R, 2005, *Everday Water: Values, Practices and Interactions*, final report by University of Western Sydney presented to Delfin Lend Lease, Sydney, NSW

Soulsby C, Gibbons CN and Robins T, 1999, Inter-basin water transfers and drought management in the Kielder/Derwent system, *Water and Environment Journal*, 13: 213–23

South East Water, 2016, *The Urban Water Cycle*, available online at www.education-southeastwater.com.au/resources, last accessed 23 March 2016

Southerton D, Chappells H and Van Vliet B (eds), 2004, *Sustainable Consumption*, Cheltenham: Edward Elgar

Speight V, 2015, Innovation in the water utilities: barriers and opportunities for US and UK utilities, *WIREs Water*, 2: 301–13, doi: 10.1002/wat2.1082

Spurling N, McMeekin A, Shove E, Southerton D and Welch D, 2013, *Interventions in Practice: Re-framing Policy Approaches to Consumer Behaviour*, Sustainable Practices Research Group, available online at www.sprg.ac.uk, last accessed 28 December 2013

Stationery Office, 1999, *Water Supply (Water Fittings) Regulations: Statutory Instrument 1999, No. 1148*, available online at www.legislation.gov.uk/uksi/1999/1148/contents/made, last accessed 7 April 2016

Stirling A, 2008, Opening up or closing down: analysis, participation and power in the social appraisal of technology, *Science, Technology and Human Values*, 33 (2): 262–94

Stockholm Environment Institute, 2011, *Understanding the Nexus: Background Paper for the Bonn 2011 Nexus Conference*, available online at http://wef-conference.gwsp.org/fileadmin/documents_news/understanding_the_nexus.pdf, last accessed 26 March 2016

Stone P, 2011, *The New River*, London Historians, available online at www.google.co.uk/?gws_rd=ssl#q=Peter+Stone+New+River, last accessed 9 April 2016

Strang V, 2004, *The Meaning of Water*, Oxford: Berg

Strengers Y, 2010, Air-conditioning Australian households: the impact of dynamic peak pricing, *Energy Policy*, 38: 7312–22

Strengers Y and Maller C, 2012, Materialising energy and water resources in everyday practices: insights for securing supply systems, *Global Environmental Change*, 22: 754–63

Sunderland D, 1999, 'A monument to defective administration?' The London Commissions of Sewers in the early nineteenth century, *Urban History*, 26 (3): 349–72

Swyngedouw E, 2004, *Social Power and the Urbanization of Water: Flows of Power*, Oxford: Oxford University Press

Swyngedouw E, 2009, The political economy and the political ecology of the hydro-social cycle, *Journal of Contemporary Water Research and Education*, 142: 56–60

Tansey J, 2004, Risk as politics, culture as power, *Journal of Risk Research*, 7 (1): 17–32

Taylor V, Chappells H, Medd W and Trentman, F, 2009, Drought is normal: the socio-technical evolution of drought and water demand in England and Wales, 1893–2006, *Journal of Historical Geography*, 35: 568–91

Texas Water Development Board, 2015, *Water Reuse*, 10 (15), available online at www.twdb.texas.gov, last accessed 18 May 2016

Thames Water, 2014a, *Final Water Resource Management Plan 2015–2040 Main Report, Section 7: Appraisal of Options*, available online at www.thameswater.co.uk, last accessed 23 June 2016

Thames Water, 2014b, *Final Water Resource Management Plan 2015–2040 Executive Summary*, available online at www.thameswater.co.uk, last accessed 23 June 2016

Thames Water, 2016a, *Promoting Metering*, available online at www.thameswater.co.uk, last accessed 23 June 2016

Thames Water, 2016b, *Water Efficiency Campaigns*, available online at www.thameswater.co.uk, last accessed 23 June 2016

Thompson M, 2008, *Organising and Disorganising*, Axminster: Triarchy Press

Thornley A, 1977, Theoretical perspectives on planning participation, *Progress in Planning*, 7 (1): 1–57

Tindall G, 1980, *The Fields Beneath*, London: Paladin

Tomoroy L, undated, *The London Water Supply Industry and the Industrial Revolution*, available online at http://eh.net/eha/wp-content/uploads/2014/05/Tomory1.pdf, last accessed 9 April 2016

Tompkins EL and Adger WN, 2004, Does adaptive management of natural resources enhance resilience to climate change? *Ecology and Society*, 9(2): 10

Tynan N, 2002, London's private water supply, in P Seidenstat, D Haameyer and S Hakim (eds), *Reinventing Water and Wastewater Systems: Global Lessons for Improving Water Management*, New York: Wiley, pp. 341–60

UK Water, 2016, *Metering: Get the Facts*, available online at www.water.org.uk, last accessed 4 April 2016

UN Water, 2013, *Total Actual Renewable Water Resources Per Capita*, available online at www.unwater.org/statistics_KWIP.html, last accessed 18 September 2013

Urich C and Rauch W, 2014, Modelling the urban water cycle as an integrated part of the city: a review, *Water Science and Technology*, 70 (11): 1857–72, doi: 10.2166/wst.2014.363

Utility Week, 2016, Are water companies doing enough to tackle leakage? 8 January, available online at utilityweek.co.uk, last accessed 18 May 2016

Vale of the White Horse, 2016, *Local Plan 2031 Part One*, available online at https://consult.southandvale.gov.uk/, last accessed 11 May 2016

Van de Brugge R and Rotmans J, 2007, Towards transition management of European water resources, *Water Resource Management*, 21: 249–67, doi: 10.1007/s11269-006-9052-0

Van der Steen P and Howe C, 2009, Managing water in the city of the future; strategic planning and science, *Reviews in Environmental Science and Biotechnology*, 8: 115–20, doi: 10.1007/s11157-009-9154-2

Van Kersbergen K and Van Waarden F, 2004, 'Governance' as a bridge between disciplines: cross-disciplinary inspiration regarding shifts in governance and problems of governability, accountability and legitimacy, *European Journal of Political Research*, 43: 143–71

Van Lieshout C, 2013, London's changing waterscapes: the management of water in eighteenth century London, unpublished thesis, Kings College London, available online at https://kclpure.kcl.ac.uk/portal/, last accessed 17 January 2017

Voß JP and Bornemann B, 2011, The politics of reflexive governance: challenges for designing adaptive management and transition management, *Ecology and Society*, 16 (2), 9

Voß JP, Daunkecht D and Kemp R, 2006, *Reflexive Governance for Sustainable Development*, Cheltenham: Edward Elgar

Wagenaar H, 2011, *Meaning in Action: Interpretation and Dialogue in Policy Analysis*, New York and London: ME Sharpe

Walker G, 2013, A critical examination of models and projections of demand in water utility resource planning in England and Wales, *International Journal of Water Resources Development*, 29 (3): 352–72, doi: 10.1080/07900627.2012.721679

Walker G, 2014, Water scarcity in England and Wales as a failure of (meta)governance. *Water Alternatives*, 7 (2): 388–413

Walker G, Cass N, Burningham K and Barnett J, 2010, Renewable energy and socio-technical change: imagined subjectivities of the 'the public' and their implications, *Environment and Planning A*, 42: 931–47

Warde A, 2005, Consumption and theories of practice, *Journal of Consumer Culture*, 5(2): 131–53

Water in the West, 2013, *Water and Energy Nexus: A Literature Review*, Stanford University, available online at http://waterinthewest.stanford.edu/, last accessed 25 September 2013

Waterwise, 2013a, *How You Use Water in Your Home*, available online at www.waterwise.org.uk/, last accessed 12 December 2013

Waterwise, 2013b, *Water: The Facts*, available online at www.waterwise.org.uk/, last accessed 28 October 2013

Wesselink A, 2007, Flood safety in the Netherlands: the Dutch response to Hurricane Katrina, *Technology in Society*, 29: 239–47

Wesselink A, 2016, *Should (Could) the UK go Dutch?* presentation to the Sustainable Research Institute, University of Leeds, 4 July

Wesselink A, Paavola J, Fritsch O and Renn O, 2011, Rationales for public participation in environmental policy and governance: practitioners' perspectives, *Environmental and Planning A*, 43(11): 2688–3409

Westling EL, 2012, Social dimensions of ecologically driven change: the case of river restoration, unpublished doctoral thesis, University of Sheffield

Westling EL, Sharp L, Rychlewski M and Carrozza C, 2014, Developing adaptive capacity through reflexivity: lessons from collaborative research with a UK water utility, *Critical Policy Studies*, 8 (4): 427–46, doi: 10.1080/19460171.2014.957334

White I, 2010, *Water in the City: Risk, Resilience and Planning for a Sustainable Future*, London: Routledge

Whittle R, Medd W, Deeming H, Kashefi E, Mort M, Twigger Ross C, Walker G and Watson N, 2010, *After the Rain: Learning the Lessons from Flood Recovery in Hull*, final project report for 'Flood, vulnerability and urban resilience: a real-time study of local recovery following the floods of June 2007 in Hull', Lancaster: Lancaster University

Wilcox D, 1994, *The Guide to Effective Participation*, York: Joseph Rowntree Trust, available online at www.partnerships.org.uk/guide/index.htm, last accessed 19 September 2012

Wittfogel K, 1957, *Oriental Despotism: A Comparative Study of Total Power*, New Haven: Yale University Press

Woelfle-Erskine C, 2015, Thinking with salmon about rain tanks: commons as inter-actions, *Local Environment*, 20 (5): 581–99, doi: 10.1080/13549839.2014.969212

Wong T and Brown R, 2008, *Transitioning to Water Sensitive Cities: Ensuring Resilience Through a New Hydro-Social Contract*, 11th International Conference on Urban Drainage, Edinburgh, Scotland

Worsley L, 2011, *If Walls Could Talk*, London: Faber and Faber

WRAP (Waste Resources and Action Programme), 2013, *Water Efficiency: The Water Label*, available online at www.wrap.org.uk/content/water-efficiency-water-label, last accessed 31 October, 2013

WRSE (Water Resources in the South East), 2013, *Progress Towards a Shared Water Resources Strategy in the South East of England, Phase 2B Report*, available online at www.wrse.org.uk, last accessed 9 May 2016

WWF (Worldwide Fund for Nature), 2009, *Interbasin Water Transfers and Water Scarcity in a Changing World: A Solution or a Pipedream?* Frankfurt: WWF

Yanow D and Schwartz-Shea P, 2006, Introduction, in D Yanow and P Schwartz-Shea (eds), *Interpretation and Method: Empirical Research Methods and the Interpretative Turn*, New York: ME Sharpe, pp. xi–xxvii

Yearley S, 2006, Bridging the science – policy divide in urban air-quality management: evaluating ways to make models more robust through public engagement, *Environment and Planning C: Government and Policy*, 24 (5): 701–14

INDEX